Matthias Gladt

Automatic reduction of multi-zone models for building simulation

Matthias Gladt

Automatic reduction of multi-zone models for building simulation

Südwestdeutscher Verlag für Hochschulschriften

Impressum / Imprint
Bibliografische Information der Deutschen Nationalbibliothek: Die Deutsche Nationalbibliothek verzeichnet diese Publikation in der Deutschen Nationalbibliografie; detaillierte bibliografische Daten sind im Internet über http://dnb.d-nb.de abrufbar.
Alle in diesem Buch genannten Marken und Produktnamen unterliegen warenzeichen-, marken- oder patentrechtlichem Schutz bzw. sind Warenzeichen oder eingetragene Warenzeichen der jeweiligen Inhaber. Die Wiedergabe von Marken, Produktnamen, Gebrauchsnamen, Handelsnamen, Warenbezeichnungen u.s.w. in diesem Werk berechtigt auch ohne besondere Kennzeichnung nicht zu der Annahme, dass solche Namen im Sinne der Warenzeichen- und Markenschutzgesetzgebung als frei zu betrachten wären und daher von jedermann benutzt werden dürften.

Bibliographic information published by the Deutsche Nationalbibliothek: The Deutsche Nationalbibliothek lists this publication in the Deutsche Nationalbibliografie; detailed bibliographic data are available in the Internet at http://dnb.d-nb.de.
Any brand names and product names mentioned in this book are subject to trademark, brand or patent protection and are trademarks or registered trademarks of their respective holders. The use of brand names, product names, common names, trade names, product descriptions etc. even without a particular marking in this work is in no way to be construed to mean that such names may be regarded as unrestricted in respect of trademark and brand protection legislation and could thus be used by anyone.

Coverbild / Cover image: www.ingimage.com

Verlag / Publisher:
Südwestdeutscher Verlag für Hochschulschriften
ist ein Imprint der / is a trademark of
OmniScriptum GmbH & Co. KG
Heinrich-Böcking-Str. 6-8, 66121 Saarbrücken, Deutschland / Germany
Email: info@svh-verlag.de

Herstellung: siehe letzte Seite /
Printed at: see last page
ISBN: 978-3-8381-3672-1

Zugl. / Approved by: Wien, TU, Diss., 2014

Copyright © 2015 OmniScriptum GmbH & Co. KG
Alle Rechte vorbehalten. / All rights reserved. Saarbrücken 2015

Abstract

Thermal simulations are an important means to predict the energy performance of buildings or groups of buildings or even whole cities. In particular with long term simulations the performance is vital for achieving good results in an acceptable amount of time.

RC models are well suited for simulating multi-zone models or running long-term simulations or a combination of both. They also have advantages regarding the clear mapping between zones and construction elements on the one side and the elements of the underlying equations on the other side. This is also helpful in the creation of an intuitive user interface because the geometric and physical properties of a building are usually better understood by users than the relevant differential equation system.

The implementation of the equation system is based on Python due to its object-orientated design as well as due to the solvers that come with Python's SciPy and due to its powerful algebraic capabilities.

Control systems for HVAC are integrated with the model at hand. Time-driven strategies are taken into account as well as temperature-driven ones in order to reflect a realistic behavior of the model. An explicit representation of the heating and cooling load is given in order to achieve accurate results without performance losses.

Despite the limited order of RC models e.g. compared with common finite-volume ones the performance may be insufficient for running multi-zone long-term simulations in an acceptable amount of time. Thus reasonable reduction algorithms are desirable in order to enhance the performance even more. Several reduction algorithms are outlined and can be combined or applied separately arbitrarily.

An important part of the algorithm regards the merging of walls depending on their physical properties. Wall merging can be applied in many cases because usually RC models are based on CAD or 3D models and the relevant reduction potential is not exploited at set-up time.

Automatic model reduction based on zone-merging can aim at accurate energy calculation or accurate calculation of temperature profiles. In both cases the reduction is based on the comparison of temperature profiles of individual zones. Influences of initial value conditions are avoided by calculating the periodic steady state prior to zone merging.

Manual overriding of automatic reduction methods is possible at all times. Thus also distinct excitations that run the risk of spoiling the simulation result can be taken into account.

Several architectures for the implementation have been investigated and a client-server architecture has been suggested. It has advantages among others regarding the adjustment of the underlying hardware with the simulation engine and the distribution of a tool and the relevant updates to the public.

The user interface is of particular interest if software is to be published. It is relevant for the developers and even more relevant for the intended users. Possible approaches range from browser-based user interfaces to handmade GUIs. An MS Excel-based user interface is presented where the implementation of the client-server communication is accomplished in a plugin.

Zusammenfassung

Gebäudesimulationen zur Vorhersage von Temperaturverläufen in Gebäuden haben in den letzten Jahren an Bedeutung zugenommen. Zunehmend wird nicht nur der Temperaturverlauf in einem einzelnen Gebäude sondern auch in Gebäudegruppen oder ganzen Städten unter Berücksichtigung der jeweiligen Wechselwirkungen simuliert.

RC Modelle zählen zu den Modelltypen die für Langzeit- und Multizonensimulationen am häufigsten verwendet werden. Neben den Vorteilen hinsichtlich der Rechenzeit verfügen sie auch über eine klare Zuordnung von Gebäudeteilen und Zonen zu den einzelnen Gleichungen des zugrundeliegenden Differentialgleichungssystems. Alle Parameter und Teile können im Gleichungssystem identifiziert werden und haben im Rahmen der Simulation eine klar definierte physikalische Bedeutung. Der eindeutige Zusammenhang zwischen dem Modell und dem dazugehörigen Gleichungssystem kann auch das Design von intuitiv zu bedienenden User Interfaces erleichtern.

Die Lösung des Gleichungssystems kann durch einen proprietären oder einen vorgefertigten Gleichungslöser erfolgen. In der vorliegenden Dissertation wurde auf Python und seine SciPy Bibliothek zurückgegriffen. Diese Bibliothek enthält leistungsfähige Implementierungen für algebraische Operationen und Gleichungslöser.

Die Bedeutung von Mess-, Steuerungs- und Regelungstechnik hat in den letzten Jahren ebenfalls zugenommen. Sowohl zeit- als auch temperaturgesteuerte Regelungen oder eine Kombination aus beiden können heutzutage problemlos umgesetzt werden. Eine Möglichkeit für die Berechnung der idealen Heiz- und Kühllasten bzw. der Heiz- und Kühlenergie wird ebenfalls in der vorliegenden Arbeit präsentiert.

Trotz der im Vergleich z.B. zu Finiten-Differenzen-Modellen üblicherweise eher geringen Anzahl an Gleichungen müssen selbst RC Modelle gelegentlich vereinfacht werden um eine Langzeitsimulationen von Multizonenmodellen in absehbarer Zeit zu ermöglichen. Daher sind Methoden um Vereinfachungen zu erzielen nach wie vor gefragt und Gegenstand der Forschung. Mehrere Möglichkeiten zur Vereinfachung werden in der gegenständlichen Arbeit präsentiert und können gemeinsam und einzeln angewendet werden.

Die Zusammenfassung mehrerer Wände in eine einzelne Wand kann bei gleichbleibender Genauigkeit der Ergebnisse zu signifikanten Verkürzungen der Laufzeit führen.

Zusammenfassung

Automatische Modellreduktionen können entweder auf eine schnelle Berechnung von Energieverbräuchen oder von Temperaturverläufen abzielen. In beiden Fällen hängt die Entscheidung über die Zusammenlegung von Zonen von der Ähnlichkeit der Temperaturverläufe im periodisch-stationären Zustand ab. Die Berechnung des periodisch-stationären Zustands ermöglicht die Elimination von Einflüssen durch Anfangsbedingungen auf das Temperaturverhalten einzelner Zonen.

Die automatische Zusammenlegung von Zonen kann jederzeit auf einfache Weise manuell übersteuert werden wodurch die Eigenheiten einzelne Modelle (z.B. durch stark unterschiedliche physikalische Eigenschaften einzelner Zonen) berücksichtigt werden können.

Mehrere Architekturen für die Implementierung der Anwendung werden untersucht und eine Client-Server Architektur als die am besten für eine Umsetzung Geeignete vorgeschlagen. Eine Client-Server Architektur hat u.a. Vorteile was die Verbreitung des Tools und die Abstimmung der Hardware mit der Implementierung betrifft.

Das User Interface ist von besonderer Bedeutung wenn die Verteilung der Software an einen größeren Benutzerkreis angedacht ist. Die Möglichkeiten reichen von browserbasierten Lösungen bis zu proprietären GUIs.

Eine Lösung in Form eines Plug-Ins für MS Excel wird vorgeschlagen, sodass die Vorteile eines Tabellenkalkulationsprogramms mit denen einer Bausimulationssoftware kombiniert werden können.

Acknowledgements

I would like to thank my supervisor Ao. Univ.-Prof. Dipl.-Ing. Dr. Thomas Bednar for supporting me as efficiently as he did during the whole work. Without his support this thesis would never have been completed. I would also like to thank Ao. Univ.-Prof. Dipl.-Ing. Dr. Ewa Weinmüller for the critical review of the thesis at hand.

Table of Contents

Abstract 1

Zusammenfassung 1

Acknowledgements 1

1   Problem Description 1
1.1   Introduction 1
1.2   Motivation and Problem Definition 2
1.3   Organization of this thesis 4

2   Setting up the model 7
2.1   State of the art 7
2.2   Modeling zones and construction elements 8
    2.2.1   Zone coupling 11
    2.2.2   Internal walls 12
    2.2.3   Interior wall under periodic conditions 13
2.3   Windows 13
2.4   Solar gains 17
    2.4.1   Shadow casting 17
        2.4.1.1   Traditional approaches 18
        2.4.1.2   GPU-based approach 20
        2.4.1.3   Conclusion 21
2.5   Modeling meteorological conditions 22

3   Implementation 23
3.1   Choosing the programming platform 23
3.2   Transferring the model into code 27
3.3   Preprocessing 30
3.4   From CAD to RC models 31

3.5	**Heat capacity values**	**33**
3.6	**Solving the equation system**	**36**
3.6.1	State of the art in building simulation	36
3.6.2	Step size	37
3.6.3	One-step vs. multi-step	38
3.6.4	Implicit vs. explicit	39
3.6.5	The choice of a solver	39
3.6.6	Assessing different solvers	40
3.6.6.1	Accuracy vs. step size	41
3.6.6.2	Performance evaluation	43
3.6.7	Conclusion	44
3.7	**Control systems**	**46**
3.7.1	State of the art, MPC	46
3.7.2	Ventilation	47
3.7.2.1	Natural ventilation	47
3.7.2.2	Mechanical ventilation	49
3.7.3	Shading	50
3.7.4	Internal gains, heating and cooling	51
3.7.5	Conclusion	58
4	**Reduction algorithm**	**60**
4.1	**State of the art in performance tuning**	**60**
4.2	**Introduction**	**62**
4.3	**Wall merging-based model reduction**	**64**
4.3.1	Wall merging	65
4.4	**Calculating the periodic state**	**66**
4.5	**Zone merging**	**67**
4.6	**Model reduction for energy calculations**	**67**

4.7	**Model reduction for zone temperature calculations**	**69**
4.8	**Results**	**70**
4.8.1	Free-float model, windows partly openable	72
4.8.2	Ideal heating and cooling systems, windows closed	75
4.8.3	Ideal heating, openable windows	78
4.9	**Conclusion**	**80**
5	**Architecture of the implementation**	**83**
5.1	**Computing power**	**87**
5.2	**Distribution and installation process**	**88**
5.3	**Network traffic**	**89**
5.4	**User interface**	**89**
5.4.1	Requirements for user-friendliness	89
	5.4.1.1 Architecture driven considerations	90
5.4.2	Browser	91
5.4.3	Handmade GUI	91
5.4.4	MS Excel	91
6	**Conclusion**	**94**
6.1	**Model**	**94**
6.2	**Implementation**	**95**
6.3	**Reduction algorithm**	**97**
6.4	**Architecture of the implementation**	**98**
6.5	**Future work**	**100**
7	**References**	**102**
7.1	**Literature**	**102**
7.2	**Figures**	**106**
7.3	**Tables**	**109**
8	**Appendix**	**111**

1 Problem Description

1.1 Introduction

The demand for long term building simulations has been growing during the recent years. On the one hand the availability of computational resources has multiplied which allows the simulation not only of single buildings but also of sets of buildings or even whole cities. On the other hand particularly in densely populated areas the interaction of buildings with each other with respect to electricity, heat, air and water supply has a significant impact on the overall consumption of resources. For instance Bueno et al. (2012) investigate the impact of the Urban Heat Island effect on the energy consumption of buildings with the help of a simulation tool.

There are several ways of creating models for thermal simulations. Neural models belong to the category of so called black box models because the input and output parameters cannot be clearly mapped with physical or geometric properties of the underlying buildings or plants. (Kramer et al., 2012) They can be well suited to describing the behavior of a building in terms of energy consumption (e.g. Neto and Fiorelli, 2008) but the model setup requires sufficient historical performance data. (Zhao and Magoulès, 2012)

Finite-volume models (Clarke, 2001) reflect the reality the best possible way but the computational effort for solving them is considerable and reasonable results for long term simulations can often not be achieved in an acceptable amount of time.

Lumped capacitance models are also called RC (resistance-capacitance) models. These models are also known as white box models. They originate from an electrical network analogy (Kramer et al., 2012) and reflect the heat flows within a building in a simplified way. The heat transfer within the individual parts of a building is modeled one-dimensional, the temperature of a construction element or a zone is considered uniform in space and the heat capacity is aggregated at a single point. Due to these assumptions RC models remain manageable even if large buildings are modeled. They can be easily set up because the individual parts of a building are clearly mapped with the components of the model. Another advantage of RC models is that they can return reasonable results for long simulation periods in a rather short amount of time.

The fact that the thesis at hand is based on the development of a RC model is mainly motivated by the following characteristics:

- The effect of physical properties can well be observed with RC models. There are no parameters without a clear physical meaning. Thus RC models allow studies regarding the sensitivity of geometric or physical modifications of a particular model the best possible way.
- Unlike with neural network models a simulation with RC models can also be run without any relevant historical performance data. It is possible to simulate buildings that are only in the process of planning.
- Long term simulations are possible in a relatively short amount of time. (Foucquier et al., 2013) A typical simulation run is rather a matter of seconds or minutes than of days or weeks.
- The model setup can be automated more easily than with other types of models. Geometric circumstances can be read from CAD files and input values which are relevant for the model setup can be calculated based on existing material databases and known parameter values.
- The accuracy of the results achieved with sensibly configured RC models can be equivalent to those of other sophisticated building simulation tools. (Dobbs and Hencey, 2012)
- Kramer et al. (2012) name user-friendliness and straight-forwardness as major advantages of RC models. The user-friendliness concerns the clear mapping of construction elements with different parts of the equation system.
- RC models open up possibilities regarding almost any kind of user interface. User interfaces can be designed in any thinkable way and there are no restrictions to text files or complicated programming interfaces.
- The integration with existing applications does usually not meet any unexpected problems because it is relatively simple to provide an easy to use interface for and RC-based simulator.
- The simulation of RC-models is based on ordinary ODE solvers. These types of solvers are very common and are included with popular math programs as well as with open source libraries for programming languages like C, Fortran or Python.

1.2 Motivation and Problem Definition

Even though RC models are suitable for fast simulations multi-zone buildings can quickly outweigh their performance advantages. For instance with a *3R2C* configuration of the walls as it is suggested in some research papers (e.g. Gouda et al., 2002, Dobbs and Hencey, 2012) a single zone with four walls, a ceiling and a floor would result into an order 13 differential equation system. Assuming the order of the equation system to be in general about ten times as high as the number of zones it becomes clear that also models of rather simple buildings quickly result into systems of several hundred or even several thousand differential equations.

Long term simulations in the sense of simulating several months or whole years cannot be completed within seconds or a couple of minutes for this magnitude of models. Thus also with RC models the longing for high performance models has rather grown than been satisfied during the recent years.

Research about model reduction focuses mainly on the reduction of order. (Kramer et al., 2012) There have been quite some efforts so far to find a simple and yet efficient reduction algorithm for RC models. In fact RC models are also appropriate for creating small models for large buildings from scratch. Achterbosch et al. (1985) created a single capacitance node for several parts of the building. They assign a value to the capacitance according to a test measurement. Nielsen (2005) presents a simple tool to evaluate energy demand and indoor environment in buildings. In his approach he sets up a thermal two-node model. Hazyuk et al. (2012) create a single zone lumped capacity model to investigate the building's thermal behavior. Dobbs and Hencey (2012) base their work on another research paper about the aggregation of states. (Deng et al., 2010)

Not a lot of research has been made regarding the reduction based on the geometric and physical characteristics of a building. So far researchers have rather focused on finding a reduction algorithm after the equation system had been set up (E.g. Deng et al., 2010, Dobbs and Hencey, 2012) or on the investigation about the best possible mapping between the individual construction parts and the lumped thermal capacity. (E.g. Gouda et al., 2002; Nielsen, 2005).

The thesis at hand aims for closing this gap in current research work by exploiting the characteristics of RC models quoted earlier the best possible way. It will outline several ways of creating full-scale RC models as well as simplified ones from buildings at any stage of a project, including the early design phase where no historical performance data exist.

It will also come up with concepts for the architecture of an implementation, be it in a local or in a distributed environment. It will be shown that web services offer a way to host different parts of an implementation on different machines also with high-end solvers at the backend which opens up new possibilities regarding the license and update policies as well as a simple distribution mechanism of only the client part.

Last but not least the creation of a user-friendly system was one the driving forces for the research work at hand. Many simulation tools seem to work well but they simply overtax users who are not part of the development team or at least very skilled in programming and building physics. To avoid misunderstandings it shall be made clear that even the usage of a building simulation engine requires some knowledge regarding the underlying physical coherences as well as abstracting capabilities in order to create useful models and to achieve acceptable results. But the goal of creating a simulation tool that combines easy to handle interfaces and high performance persists and remains to be accomplished in the course of the thesis.

1.3 Organization of this thesis

In the following chapter the setup of the model is explained in detail. It is outlined how a multi-order RC model can be created from an existing CAD model or a simple sketch of a building. Construction elements are mapped with two capacitances and three resistances (*3R2C* model) each. The zone is assigned its own capacitance representing the thermal capacity of the room air and the furnishing.

Many models are restricted to a single zone. This is often not sufficient if a multi-zone building has to be modeled in order to reflect the energy performance well enough for further planning. Zone coupling is one of the key features of the model that is going to be presented.

Windows are often modeled as multilayered construction elements. An approach is presented where the drawbacks of a multilayered model regarding the overall order of the model are avoided by neglecting the thermal capacity of the windows which is usually very low compared with those of other construction elements. Thus the order of the model can be maintained despite taking into account the effects of windows on the temperature profile of a particular zone.

Calculating solar gains is tightly coupled with shadow mapping. The principle of a matrix-based calculation of shadow casting based on polygon clipping is presented. The approach can be fully automated and easily integrated with existing applications.

The implementation of the model is outlined in its own chapter. It includes the choice of a qualified programming language as well as of an applicable solver. The choice of a suitable solver is vital for achieving good performance when a simulation is run. The pros and cons of different solvers are given as well as a performance comparison and the most important configuration details.

The integration of control systems regarding HVAC, heating, cooling, etc. is always an issue with simulation programs. It can only be designed after the model has been set up and the concept of implementation is ready. An approach is presented that allows the simulation of temperature-controlled systems as well as of time-controlled ones.

The final part of the implementation chapter of the thesis will give application examples and results which form the basis for the core chapter of the thesis, a reduction algorithm to reach the best possible tradeoff between a loss of accuracy and a gain of performance. It will give examples of how the algorithm can be applied and what it means for the performance of building simulation.

A chapter about the architecture of the implementation of a building simulation tool focuses on the pros and cons of a distributed environment compared with a local one and on creating a user-friendly interface. License strategies have to be taken into account as well as considerations regarding a local or a distributed runtime environment if a commercial exploitation is planned.

A graphical representation of the different parts of the thesis is given in Figure 1.

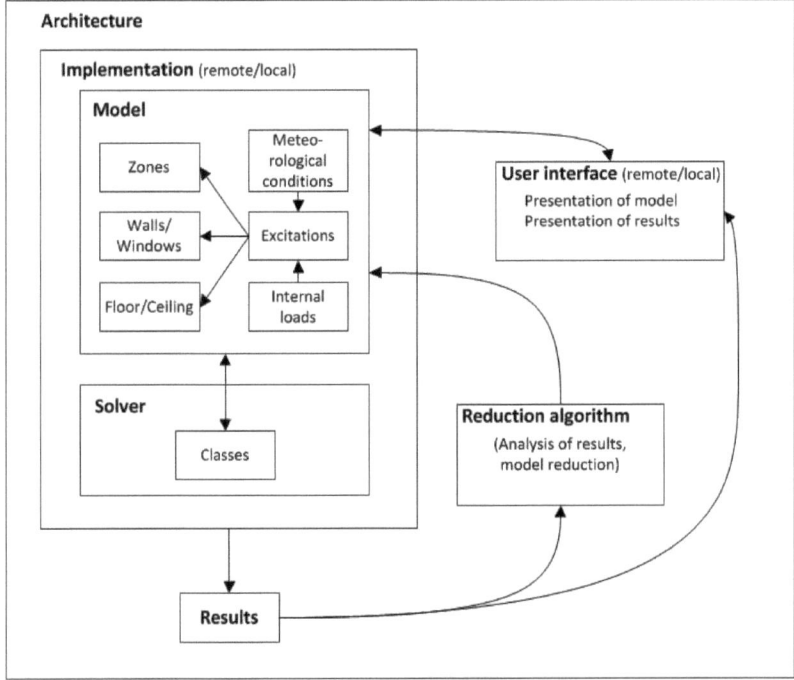

Figure 1: Graphical presentation of the organization of the thesis at hand.

2 Setting up the model

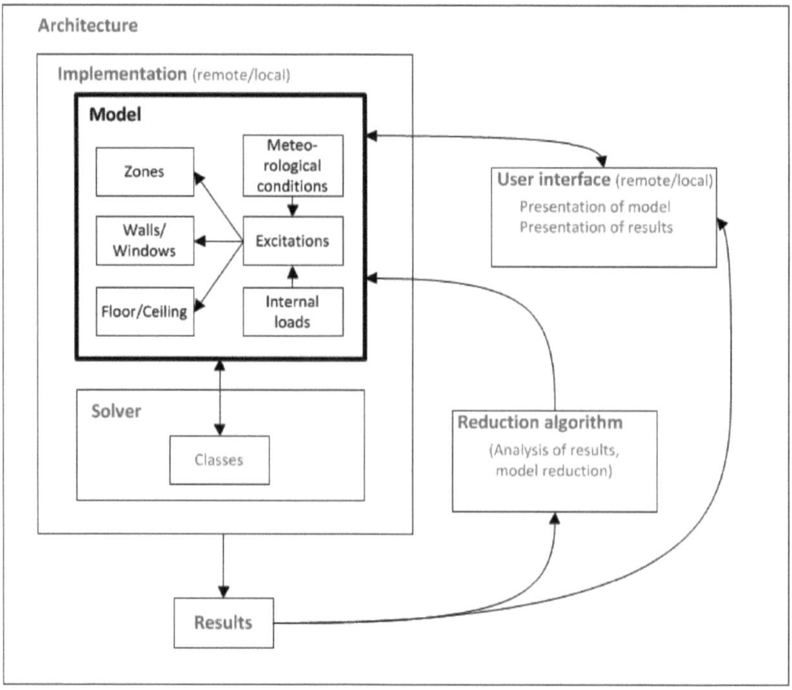

Figure 2: In the context of the thesis the model represents a central part.

Figure 2 displays the part of the model in the context of the whole thesis. The model consists of several sub models including the building, construction elements and the windows associated with them was well as the excitations by meteorological conditions and internal loads.

The individual sub models are explained in detail in the following.

2.1 State of the art

The decision has been made to base further research work on a RC model. Two types of nodes are created for setting up the model: those associated with zones and those with construction elements. There has been a lot of research regarding the setup of RC models and in particular the modeling of construction elements. At all times the goal was to keep the number of nodes representing construction elements low and to provide sufficient accuracy of the simulation at the same time. As already mentioned Achterbosch (1985) was among the first to deal with RC models. He created a lumped capacity for several construction parts which he split up into two capacities according to a test measurement. Rumianowski et al. (1989) set up a *2R3C* model at the time and Gouda et al. (2002) prove that a building envelope composed of construction elements represented by *3R2C* models returns significantly more accurate result than *2R1C* – based presentations. Also Jiménez et al. (2008) use a *3R2C* model in their research paper. Like Gouda et al. (2002) they focus on determining the relevant parameters of their models rather than on finding a reduction algorithm for a model comprising multiple zones. Hazyuk et al. (2012) sets up a *2R1C* model to investigate model predictive control.

There is less diversity regarding the zonal nodes. In most research works a single capacitance is associated with the zonal node.

2.2   Modeling zones and construction elements

RC models evolved from an electrical analogy where the electrical resistance corresponds with thermal resistance, thermal capacity with electrical capacitance and the heat flow with electrical current. Accordingly a presentation of a RC model resembles one of an electrical network.

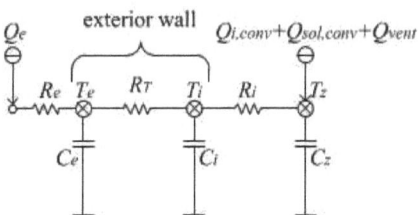

Figure 3: Presentation of a zone to which an exterior wall is attached.

The capacities C_i and C_e present the thermal capacities associated with the surfaces of the relevant construction element, in this case an exterior wall. C_Z is the internal heat capacity of the only zone, "Z". R_e and R_i represent the convective heat transfer resistances at the outer and the inner surface respectively. R_T stands for the conductive resistance within the wall. T_i and T_e denote the surface temperatures of the wall and T_Z the air temperature of the zone.

Q denotes the internal and the external excitations, respectively. In particular $Q_{i,conv}$ stands for the convective part of the total of the inner loads within a zone. The inner loads comprise the energy emitted by the persons present as well as the energy emitted by machines like computers, the copy machine, the coffee machine, etc. It may be useful to separate the energy produced by persons and the one from e.g. computers because the ratio between the radiative part and the convective one are not equal in both cases. Q_{vent} represents the total of the gains and losses by natural and mechanical ventilation and $Q_{sol,conv}$ the convective part of solar gains transmitted through transparent parts of the building.

Natural ventilation is modeled as follows:

$$q_{vent,nat} = \dot{V} \cdot c_{P,air} \cdot \rho_{air} \cdot (T_z - T_{a,e}) \qquad (2.1)$$

$c_{P,air}$ is the specific heat of air, ρ_{air} its density and $\dot{V}$ the air flow rate.

The air flow rate with natural ventilation is assumed with

$$\dot{V} = C_{ref} \cdot A \cdot \sqrt{H} \cdot \sqrt{T_z - T_{a,e}} \qquad (2.2)$$

where C_{ref} (m$^{0.5}$ h$^{-0.5}$ K$^{-0.5}$) represents the exchange coefficient.

Q_e represents the sum of the convective and radiative excitations caused from outside through the meteorological conditions.

The relevant equation system can be set up in a continuous form or in a discrete-time form. The discrete-time form is appropriate if the time step needs to be chosen explicitly. The continuous form allows the application of a solver where the time step is chosen dynamically.

The continuous form reads:

$$C_e \frac{dT_e}{dt} = \frac{\dot{Q}_e(t)}{R_e} + \frac{T_i - T_e}{R_T} \qquad (2.3)$$

$$C_i \frac{dT_i}{dt} = \frac{T_e - T_i}{R_T} + \frac{T_z - T_i}{R_i} + \frac{T_{Rad,Z} - T_i}{R_{Rad,Z}} \qquad (2.4)$$

$$C_Z \frac{dT_Z}{dt} = Q_{i,conv,Z}(t) + Q_{sol,conv,Z}(t) + Q_{vent,Z}(t) + \frac{T_i - T_Z}{R_i} \qquad (2.5)$$

A state-space presentation of the equations 2.3-2.5 reads

$$\frac{d\vec{T}}{dt} = \tilde{A} \cdot \vec{T} + \tilde{B} \cdot \vec{U} \qquad (2.6)$$

where $\vec{T}$ represents the state vector, i.e. a vector of zone and surface temperatures, $\tilde{A}$ and $\tilde{B}$ are coefficient matrices according to the equations 2.3-2.5 and $\vec{U}$ a vector representing the excitations of the system. Due to its compactness a state space presentation of the system is often used with algebraic math tools or third party equation solvers.

$T_{Rad,Z}$ is the "radiative" temperature of zone "Z" which is defined as:

$$T_{Rad,Z} = \frac{\sum_j A_j \cdot T_{i,j} + \sum_k A_k \cdot T_{w,i,k} + Q_{Rad,Z}(t)}{\sum_j A_j + \sum_k A_k} \qquad (2.7)$$

$Q_{Rad,Z}$ is the total of the radiative amount of all loads, i.e. inner loads by persons and machines and solar gains through transparent parts of the building envelope.

A_j denotes the surface area of construction element j and $T_{i,j}$ denotes the temperature at the inner surface of construction element j.

A_k denotes the surface area of window k and $T_{w,i,k}$ denotes the temperature at the inner surface of window k. The calculation of $T_{w,i,k}$ will be treated in the following chapter.

The discrete-time form presentation is useful if for instance frequent internal or external load changes cause discontinuities in the load profiles which accordingly lead to frequent changes of the curvature of the temperature profile. Solvers of a continuous system may not be able to perceive all these changes, in particular if the points where the loads change are not known in advance or if the solver cannot be synchronized with the discontinuities.

The discrete-time form of the above equation based on the time step Δt reads:

$$C_e \cdot \Delta T_e = \left(\frac{Q_e(t)}{R_e} + \frac{T_i - T_e}{R_T} \right) \cdot \Delta t \qquad (2.8)$$

$$C_i \cdot \Delta T_i = \left(\frac{T_e - T_i}{R_T} + \frac{T_Z - T_i}{R_i} + \frac{T_{Rad,Z} - T_i}{R_{Rad,Z}} \right) \cdot \Delta t \qquad (2.9)$$

$$C_Z \cdot \Delta T_Z = \left(Q_{i,conv,Z}(t) + Q_{sol,conv,Z}(t) + Q_{vent,Z}(t) + \frac{T_i - T_Z}{R_i} \right) \cdot \Delta t \qquad (2.10)$$

Figure 3 shows a presentation of a very simple model. Thinking of a single zone model as a room with four walls, a floor and a ceiling, five construction elements would have to be added, all exterior ones, one, the floor, adjoining the ground.

2.2.1 Zone coupling

Many research papers use single-zone models or replace multiple construction elements with a single lumped heat capacity. (E.g. Nielsen, 2005; Hazyuk et al., 2012) In the approach at hand temperature profiles of multiple zones shall be calculated. Thus the aggregation of several zones into one is not possible a priori. Instead zones are separated via interior walls which are modeled the same way as exterior ones and are specified by the same physical properties. Creating models with a theoretically unlimited number of zones each disposing of a theoretically equally unlimited number of construction elements is possible. Coupling one zone with another is an easy task with the *RC* model at hand.

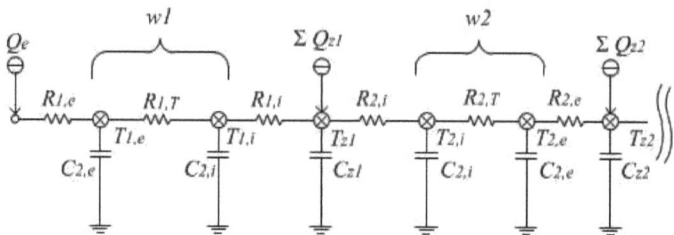

Figure 4: Presentation of a two-zone model to which an exterior wall is attached. The cut on the right hand side indicates that the model could be infinitely continued with further construction elements and zones.

Figure 4 displays a two-zone model where the partition wall is represented by the wall, "*w2*". Whereas the differential equation representing the change of zone temperature is similar as eq. 2.6 the equations for the change of the surface temperatures of the partition wall read:

$$C_{2,i} \frac{dT_{2,i}}{dt} = \frac{T_{2,e} - T_{2,i}}{R_{2,T}} + \frac{T_{Z1} - T_{2,i}}{R_{2,i}} + \frac{T_{Rad,Z1} - T_{2,i}}{R_{Rad,Z1}} \qquad (2.11)$$

$$C_{2,e} \frac{dT_{2,e}}{dt} = \frac{T_{2,i} - T_{2,e}}{R_{2,T}} + \frac{T_{Z2} - T_{2,e}}{R_{2,e}} + \frac{T_{Rad,Z2} - T_{2,e}}{R_{Rad,Z2}} \qquad (2.12)$$

The approach can also be applied recursively for the setup of equations associated with further construction elements or zones adjoining zone *z1* or *z2*. That way an equation system is set up holding two differential equations per construction element and one per zone.

2.2.2 Internal walls

Whereas partition walls have different zones adjoined to each surface and the equations for the change of temperature can be set up according to eq. 2.10 and eq. 2.11, internal walls are within one zone and their surfaces are adjoined to the same zone. Figure 5 gives an RC presentation of such a wall. In this case the exterior surface temperature T_e is replaced with another internal surface temperature, T_{i2}. Both surface temperature nodes, T_{i1} and T_{i2} are coupled with the zone temperature node, T_z.

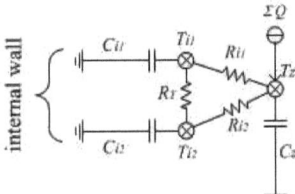

Figure 5: RC presentation of a non-symmetric internal wall. Both surface nodes are coupled with the same zone.

The relevant equations read:

$$C_{i1}\frac{dT_{i1}}{dt} = \frac{T_{i2}-T_{i1}}{R_T} + \frac{T_z-T_{i1}}{R_{i1}} + \frac{T_{Rad,Z}-T_{i1}}{R_{Rad,Z}} \quad (2.13)$$

$$C_{i2}\frac{dT_{i2}}{dt} = \frac{T_{i1}-T_{i2}}{R_T} + \frac{T_z-T_{i2}}{R_{i2}} + \frac{T_{Rad,Z}-T_{i2}}{R_{Rad,Z}} \quad (2.14)$$

In case of a symmetric internal wall both surfaces have the same temperature. Thus one of the equations can be removed from the equation system. Figure 6 holds a RC representation of a symmetric internal wall.

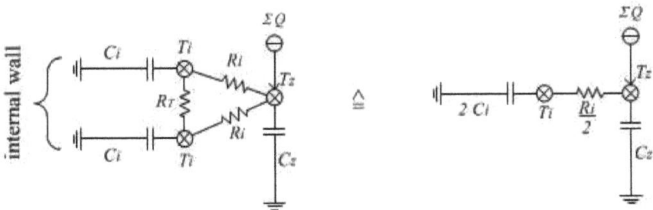

Figure 6: RC presentation of a symmetric internal wall. Since the values for the thermal capacity and resistance are the same on both surfaces, one node can be removed.

The relevant equation in this case reads:

$$C_i \frac{dT_i}{dt} = \frac{T_i - T_i}{R_T} + \frac{T_Z - T_i}{R_i} + \frac{T_{Rad,Z} - T_i}{R_{Rad,Z}} = \frac{T_Z - T_i}{R_i} + \frac{T_{Rad,Z} - T_i}{R_{Rad,Z}} \quad (2.15)$$

2.2.3 Interior wall under periodic conditions

Sometimes a model is assumed to be symmetric in relation to a particular wall when the equations for the surface temperatures are set up. This is a way to simulate many connected identical zones.

Thus zone modeling on one side of the wall can be left aside and the conditions on both sides of the wall are assumed identical. In this case also both surface temperatures of the relevant wall are identical and one of the differential equations for the surface temperatures can be removed just like for an internal wall.

2.3 Windows

The thermal properties of windows have improved to a remarkable extent during the last decades. For instance an overall heat transfer coefficient of below 1 (W m^{-2} K^{-1}) is not uncommon for modern windows nowadays. Nevertheless windows still have a significant impact on the energy performance of a building. Huang and Brodrick (2000) estimate that 32 % of the heat gains that contribute to the U.S. net national residential cooling load go back to solar gains through windows. Regarding the heating period they estimate windows to contribute about 25 % to the overall heat losses of the entire residential sector. This is about in accordance with Arasteh et al. (2006) who estimate that Windows in the U.S. consume 30 % of building heating and cooling energy.

Windows contribute to the thermal behavior of a zone in two ways: In that they have a different surface temperature than the wall that they belong to and by allowing irradiation to pass through the glazing and to contribute to the solar gains.

The calculation of the solar gains depends on the shadowed parts of the building envelope and is explained at a later point.

Whereas in many research papers about RC models the calculation of solar gains is taken into account the calculation of the surface temperatures is not. (E.g. Nielsen, 2005) The calculation of the surface temperature can be accomplished in several ways. Winkelmann (2001) examines how windows are modeled in EnergyPlus (2013): By solving the heat balance equations the glass face temperatures can be determined. The drawback of the approach is that a window with N glass layers requires $2N$ equations which have to be solved for every time-step. Pal et al. (2009) focus on the temperature profile within the glazing by setting up a finite-difference model. Freire et al. (2011) compare two window models: a simplified resistive one and finite-volume based one. In both cases they calculate the transmitted solar energy as well as the inner and the outer surface temperatures.

Hazyuk et al. (2012) state that windows do not accumulate thermal energy. This is why they need not be assigned their own thermal capacities.

The approach that was followed allows a complete separation of the calculation of the transmitted solar energy from the calculation of the temperature at the surface of the glazing and of the heat conduction through the window respectively. This is advantageous if the calculation of the solar gains is part of a different component than the actual simulation engine which means that it can be hosted on a different machine. It is even possible to complete the calculation of solar gains in the preprocessing phase and pass them on to the simulation engine at a later point.

The goal of the approach outlined in the following is the calculation of the surface temperatures of the glazing. It starts with the formulation of the density of the heat flow rate at the inner side of the window:

$$q_{w,i} = h_{c,i} \cdot (T_{w,i} - T_{a,i}) + h_r \cdot (T_{w,i} - T_{Rad,Z}) \qquad (2.16)$$

$T_{w,i}$, represents the temperature at the inner surface of a window, $T_{a,i}$, the air temperature of the relevant zone, $h_{c,i}$ the convective heat transfer coefficient, h_r, the radiative heat transfer coefficient and $T_{Rad,Z}$ the radiative temperature which is defined according to eq. 2.8. Introducing the effective heat transfer coefficient at the inner side of a window,

$$h_{eff,i} = h_{c,i} + h_{r,i} \qquad (2.17)$$

$q_{w,i}$ results to:

$$q_{w,i} = h_{eff,i} \cdot T_{w,i} - (h_{c,i} \cdot T_{a,i} + h_{r,i} \cdot T_{r,i}) \qquad (2.18)$$

$$q_{w,i} = h_{eff,i} \cdot T_{w,i} - ((h_{c,i} + h_{r,i}) \cdot T_{a,i} + h_{r,i} \cdot (T_{r,i} - T_{a,i})) \qquad (2.19)$$

$$q_{w,i} = h_{eff,i} \cdot \left(T_{w,i} - \left(T_{a,i} + \frac{h_{r,i}}{h_{eff,i}} \cdot (T_r - T_{a,i}) \right) \right) \qquad (2.20)$$

Defining an effective temperature at the inside surface of a window,

$$T_{eff,i} = T_Z + \frac{h_{r,i}}{h_{eff,i}} \cdot (T_{Rad,Z} - T_Z) \qquad (2.21)$$

$q_{w,i}$ results to:

$$q_{w,i} = h_{eff,i} \cdot (T_{w,i} - T_{eff,i}) \qquad (2.22)$$

The definition of the exterior radiative heat transfer coefficient at the outer surface, $h_{r,e}$, is based on a linearized version of the Boltzmann law which reads:

$$q_{i,j} = 4 \cdot \sigma \cdot \varepsilon \cdot \overline{T}_{i,j}^{\,3} \cdot (T_i - T_j) \qquad (2.23)$$

with $q_{i,j}$ (W m²) the density of the heat flow between surface i and surface j, σ, the Boltzmann constant, ε, the ratio between the real and the blackbody total emissive radiation. (Hagentoft 2001) $\overline{T}_{i,j}$ (K) is defined as:

$$\overline{T}_{i,j} = \frac{T_i + T_j}{2} \qquad (2.24)$$

Thus $h_{r,e}$ results to:

$$h_{r,e} = 4 \cdot \sigma \cdot \varepsilon \cdot \overline{T}_{i,j}^{\,3} \qquad (2.25)$$

The density of the heat flow rate at the exterior side of the window is formulated in a similar way as the one at the inner side of the window:

$$q_{w,e} = q_{c,e} + q_{r,e} \qquad (2.26)$$

$q_{c,e}$, denotes the convective part of the heat flow part and $q_{r,e}$ the radiative one:

$$q_{c,e} = h_{c,e} \cdot (T_{w,e} - T_{a,e}) \qquad (2.27)$$

$$q_{r,e} = h_{r,e} \cdot SVF \cdot (T_{w,e} - T_{sky}) + h_{r,e} \cdot (1 - SVF) \cdot (T_{w,e} - T_{a,e}) \qquad (2.28)$$

$T_{w,e}$ represents the temperature at the outer surface of a window, $T_{a,e}$ the outside air temperature of the relevant zone, T_{sky} the sky temperature and $h_{c,e}$ the exterior convective heat transfer coefficient. *SVF* represents the sky view factor which is defined as the relationship between the visible portion of the sky and the portion covered by the surrounding objects. Gladt and Bednar (2013b) present a method of calculating the *SVF* for scenarios made up of arbitrary polygonal faces.

Inserting the effective heat transfer coefficient at the outer surface, $h_{eff,e}$, which is defined as the sum of $h_{c,e}$ of and $h_{r,e}$,

$$h_{eff,e} = h_{c,e} + h_{r,e} \qquad (2.29)$$

$q_{w,e}$ results to

$$q_{w,e} = h_{eff,e} \cdot T_{w,e} - h_{eff,e} \cdot \left(T_{a,e} - \frac{h_{r,e}}{h_{eff,e}} \cdot SVF \cdot T_{a,e} + \frac{h_{r,e}}{h_{eff,e}} \cdot SVF \cdot T_{sky} \right) \qquad (2.30)$$

$T_{\mathit{eff},e}$ is defined in analogy to Hagentoft's equivalent exterior temperature (2001):

$$T_{\mathit{eff},e} = T_{a,e} + \frac{h_{r,e}}{h_{\mathit{eff},e}} \cdot \left(SVF \cdot T_{sky} + (1-SVF) \cdot T_{a,e} - T_{a,e} \right) \qquad (2.31)$$

Thus $q_{w,e}$ results to

$$q_{w,e} = h_{\mathit{eff},e} \cdot \left(T_{w,e} - T_{\mathit{eff},e} \right) \qquad (2.32)$$

Under stationary conditions the following is valid:

$$h_{\mathit{eff},i} \cdot \left(T_{w,i} - T_{\mathit{eff},i} \right) = \frac{1}{R_G} \cdot \left(T_{w,i} - T_{w,e} \right) \qquad (2.33)$$

$$h_{\mathit{eff},e} \cdot \left(T_{w,e} - T_{\mathit{eff},e} \right) = \frac{1}{R_G} \cdot \left(T_{w,i} - T_{w,e} \right) \qquad (2.34)$$

$$U_{\mathit{eff}} \cdot \left(T_{\mathit{eff},e} - T_{\mathit{eff},i} \right) = \frac{1}{R_G} \cdot \left(T_{w,i} - T_{w,e} \right) \qquad (2.35)$$

where R_G represents the heat transfer resistance of the glazing and U_{eff} the effective heat transition coefficient:

$$U_{\mathit{eff}} = \frac{1}{\frac{1}{h_{\mathit{eff},i}} + R_G + \frac{1}{h_{\mathit{eff},e}}} \qquad (2.36)$$

Isolating $T_{w,i}$ results to:

$$T_{w,i} = T_{\mathit{eff},i} - \frac{\frac{1}{h_{\mathit{eff},i}}}{\frac{1}{h_{\mathit{eff},i}} + R_G + \frac{1}{h_{\mathit{eff},e}}} \cdot \left(T_{\mathit{eff},i} - T_{\mathit{eff},e} \right) \qquad (2.37)$$

$T_{w,i}$ results to:

$$T_{w,i} = T_{\mathit{eff},i} - \frac{U_{\mathit{eff}}}{h_{\mathit{eff},i}} \cdot \left(T_{\mathit{eff},i} - T_{\mathit{eff},e} \right) \qquad (2.38)$$

Since $T_{Rad,Z}$ is referenced in $T_{\mathit{eff},i}$ (eq. 2.21) and $T_{\mathit{eff},i}$ in $T_{w,i}$ (eq 2.38) and $T_{w,i}$ in $T_{Rad,Z}$ (eq. 2.7) there is a circular reference.

However, since there are three variables in three equations the three of them can be expressed explicitly.

A math program that worked well with the equations mentioned before in order to reach an explicit representation of $T_{Rad,Z}$, $T_{\mathit{eff},i}$ and $T_{w,i}$ was Maxima. The explicit representation of $T_{Rad,Z}$ reads:

$$T_{Rad,Z} = \frac{\sum_k \left(\left(h_{eff,i,k} \cdot T_{eff,e,k} + \left(h_{r,i} - h_{eff,i} \right) \cdot T_{a,e} \right) \cdot U_{eff,k} + \left(h_{r,i} + h_{eff,i} \right) \cdot h_{eff,i} \right) \cdot A_k + \sum_j T_{i,j} \cdot A_j + Q_{Rad,Z}}{\sum_k \left(h_{r,i} \cdot U_{eff,k} + h_{eff,i} \cdot \left(h_{eff,i} - h_{r,i} \right) \right) \cdot h_{r,i} \cdot A_k + \sum_j T_{i,j} \cdot A_j \cdot h_{eff,i}^2 \cdot h_{r,i}}$$

(2.31)

where k denotes the index of the relevant window of a construction element belonging to zone Z and j the index of the relevant construction element of zone Z.

2.4 Solar gains

The total of irradiation adding to the loads of a specific zone depends on the location and the point of time. It is composed of direct and diffuse irradiation. Diffuse irradiation on its part is composed of diffuse irradiation originating directly from the sky and of diffuse irradiation reflected from the ground.

There are two ways of making out the actual values of irradiation:

- Using historical data from measurements or fictive weather data that was synthetically created based on real measurement data with climate prediction models. This data is often stored in "TMY" (typical meteorological year) files. TMY files contain weather data for a whole year in a text format with one line per hour. Until today there are three different formats of TMY files. The relevant file endings are .tmy, .tm2 and .tm3. TMY files are pretty common and can be found in the internet for many different locations in the world.
- Creating the data according to an irradiation model. Many models exist for the prediction of irradiation data. The Austrian standard, ÖNORM B 8110-3 (2012), contains one as well as the German standard VDI 6007 / 3 (2012), Perez et al. (1990) have created one to name a few of them.

With the research work for the thesis at hand either the model from the Austrian standard, ÖNORM B 8110-3 (2012) was used or the data from a freely available TMY file. With the data from the TMY file the conversion methods of the VDI 6007 / 3 (2012) model were used to translate the values according to the orientation and the incline of the relevant surface.

2.4.1 Shadow casting

Realistic building simulations require the allowance for the effect of shadow casting on the overall solar gains of a particular zone. The irradiation data gained from the relevant model or from existing data only contains values for completely unshadowed surfaces. Thus the values for direct solar irradiation need to be reduced according to the ratio between the shadowed and the unshadowed portion of a particular face. Accordingly Jones, Greenberg and Pratt (2011) define the projected sunlit surface fraction as the dimensionless ratio of incident beam radiation to unobstructed direct normal radiation. They take into account the angle between the sun's rays and the surface of interest in their definition of the intensity of direct radiation. If the amount of irradiation has been reduced according to the orientation and the incline of the relevant surface already before the calculation of the shadowed parts no additional reduction has to be made.

2.4.1.1 Traditional approaches

Traditional approaches comprise mainly polygon-based ones and ray tracing-based ones. The principals of both methods have been described as early as 1979. (Walton) In his work about the shadow calculation in EnergyPlus (2013) Winkelmann (2001) states that an overlapping polygon method is used. Nielsen (2005) calculates the transmitted solar energy based on Perez' model of solar radiation. (Perez et al., 1990) He focuses on the calculation of the shading by window recess and overhangs. Gladt and Bednar (2013a) present a way of calculating the shadows with the help of matrix-based coordinate transformations and polygon clipping. They organize their approach in two parts: In the first part they solve the visibility problem according to the Weiler-Artherton algorithm (Weiler and Atherton, 1977). Figure 7 displays a typical scenario that the algorithm is applied at. Only the parts within the projection of the foremost polygon are taken into account within a particular sorting phase. The algorithm is applied recursively to the remaining outer parts of all polygons when the inner part of the last polygon has been finished processing.

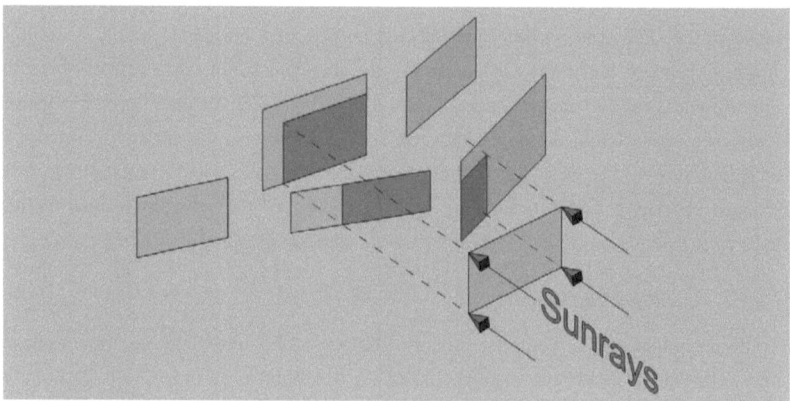

Figure 7: According to the Weiler-Artherton algorithm the parts outside of the projection of the foremost polygon are clipped. Only the inside parts are sorted. (Gladt and Bednar, 2013)

In the second part they project the polygons of interest unto a projection plane orthogonal to the sun. After rotating the projection plane into a 2D coordinate system they apply a polygon clipping library for calculating the shadowed parts based on the order set up in the first step. (Figure 8)

For the details of the matrix operations, the implementation of the algorithm and a comparison with a GPU-based approach refer to Gladt and Bednar (2013a).

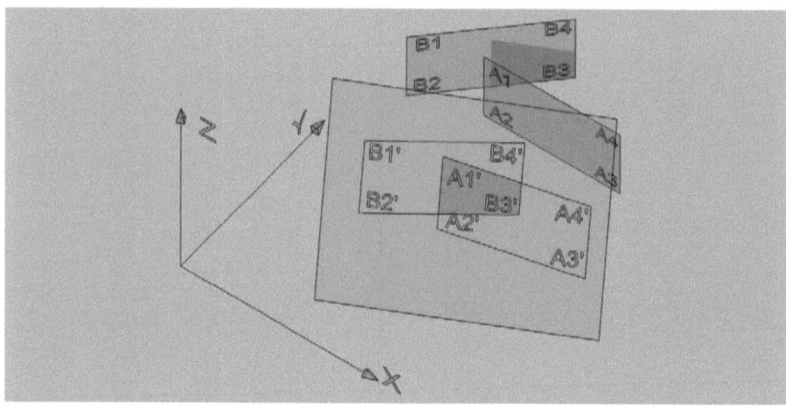

Figure 8: Polygon clipping is applied after the relevant faces have been projected unto a plane orthogonal to the sun. (Gladt and Bednar, 2013)

2.4.1.2 GPU-based approach

GPU-based shadow mapping is based on pixel counting. Recently GPU-based calculations of shadow mapping have become more and more popular. Jones et al. (2012) set up a GPU-based implementation for a fast calculation of the shadowed parts of complicated building envelopes. GPU-based shadow mapping has several pros and cons some of which will be named in the following:

Pros:

- GPUs can help to complete shadow casting calculations in real time. Also complicate scenarios with several thousands of polygons can be processed in milliseconds.
- If the implementation is set up in a sensible way the accuracy of the results comes close to the one of traditional approaches where the accuracy is only limited by numerical imprecisions.
- A graphical presentation of the scenario being processed can easily be created since GPUs are built to generate the relevant picture and to render it on the screen. This can be helpful with plausibility tests and for presentation purposes.

Cons:

- GPUs have a rather complicated programming interface. They are designed as state machines which have to be accessed in a way that many programmers are unfamiliar with.
- Various matrices have to be set up for every run of pixel counting and for every polygon of interest anew. The set up process is programmed only once but in complicated scenarios there is no guarantee that all potential geometric complications are taken into account.
- The viewport has to be set either once for the whole scene or once for every polygon of interest. If it is set once per scenario the accuracy of the results may suffer according to the size of the viewport. If it is set once per polygon it has to be set anew for every polygon. However also in this case the accuracy of the results may also suffer e.g. if the minimum and the maximum coordinates of the transformed geometry require a large viewport with respect to the polygon of interest.
- Accuracy in general can be an issue with pixel counting. A certain number of pixels are mapped with a particular polygon. If the number of pixels is too low the shadowed parts cannot be precisely calculated.

2.4.1.3 Conclusion

The goal of the implementation at hand was to create an easy to use interface that takes scenarios consisting of an arbitrary number of polygons and the vector of the sunrays.

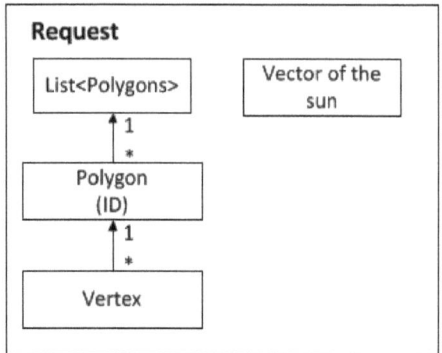
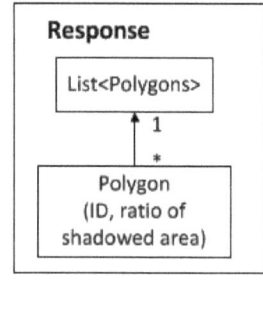

Figure 9: Schematic representation of the objects passed to and returned from the interface of the implementation.0

Figure 9 displays a way to setup of the request and the response object of the interface of the relevant component. Since only a vector representation of all parts of the building envelope and the surrounding objects are needed both objects can be kept very simple. The payload of the response object simply consists of a list of polygon objects, each holding the ID of the relevant polygon object from the request and the ratio of the shadowed area. The IDs are needed to map the values of the response list with the polygons from the request. The number of vertices per polygon is unlimited.

A component implemented for testing purposes was based on the Weiler-Artherton algorithm described earlier. It proved to perform well and reliably. The execution time was below a minute for several thousands of polygons. Considering that the time for running a simulation of a building with an envelope composed of several thousand walls exceeds the execution time of shadow calculation by some magnitudes the performance of the traditional approach seems acceptable.

In case of a bottleneck at the shadow calculation component a GPU-based implementation can be set up to improve the performance.

2.5 Modeling meteorological conditions

The exterior excitations in the model displayed in Figure 3 are represented by Q_e. Q_e represents the cumulated excitations on the model through the outside temperature and through solar gains and losses.

The impact of Q_e on the building model can be replaced with an equivalent exterior temperature, T_{eq}. Hagentoft (2001) associates the equivalent exterior temperature with the surface of a construction element. Kämpf and Robinson (2007) follow the approach in that they add a link between the wall temperature node of their model and the external temperature node. Taking into account the outside temperature and the radiative interaction of the building with the atmosphere T_{eq} can be expressed as:

$$T_{eq} = \frac{1}{h_{eff}} \cdot \left(\frac{Q_{sol,opaque}}{A_w} + T_{a,e} \cdot h_c + T_{sky} \cdot h_r \right) \quad (2.16)$$

$Q_{sol,opaque}$ represents the solar gains absorbed by the opaque part of the relevant construction element, and A_w the area. Another representation of T_{eq} reads:

$$T_{eq} = T_{a,e} + \frac{1}{h_{eff}} \cdot \left(\frac{Q_{sol,opaque}}{A_w} + h_r \cdot (T_{sky} - T_{a,e}) \right) \quad (2.17)$$

A major advantage of the presentation of the meteorological excitations by T_{eq} is that all relevant calculations can be accomplished in the pre-processing phase of the building simulations. An extra component with an interface to transform weather files into equivalent exterior temperatures can be set up which avoids any negative performance impact on the actual simulation engine.

3 Implementation

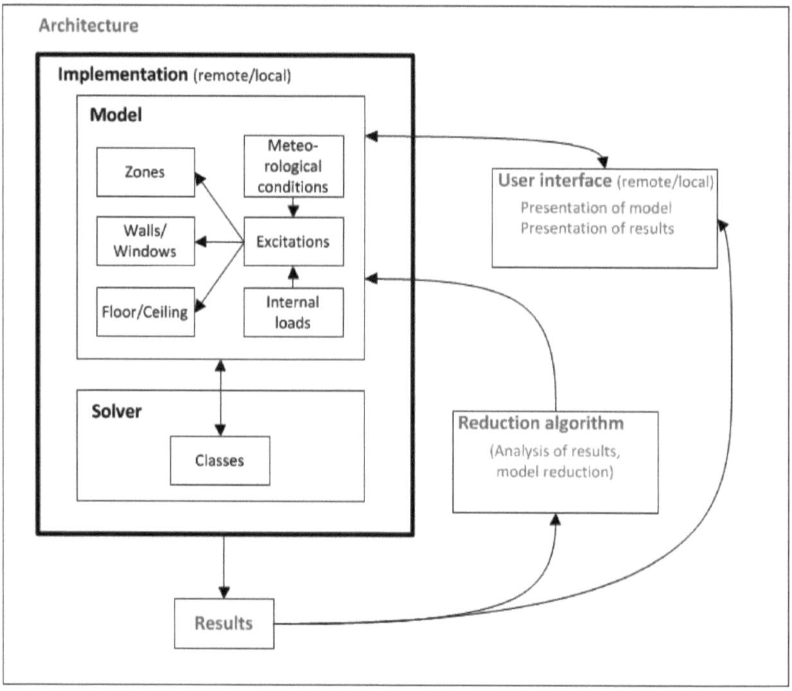

Figure 10: In the context of the thesis the implementation concerns the model and the solver.

Figure 10 displays the implementation in the context of the thesis at hand. It includes not only the model as it was explained in 2.2 and 2.3 but also the excitations by meteorological conditions and internal loads.

3.1 Choosing the programming platform

The implementation of a model like the one introduced above requires a programming platform capable of fulfilling certain requirements. Some of these requirements are mandatory and some are only nice to have. In the following the most important ones are summarized in order to provide a decision-making basis.

- The programming platform has to support the derivation of the equation system from the model of a building. The setup of the equation system shall be implemented based on an object-orientated representation of the building. Loops and conditional statements must be possible in order to automate the setup of the equation system independently of the size of the building.
- An object-orientated language is useful. A simple mapping between the individual construction elements and zones with instances of the relevant classes can easily be set up which facilitates an intuitive handling of the building models.
- As already outlined earlier the model results into a differential equation system for which a solver is needed. The integration of a third-party solver should be possible via an applicable binding or by accessing it directly within the relevant programming platform. If no third-party solver shall be used a proprietary one has to be written.
- If a third-party solver is used it has to return reliable results and dispose of an acceptable performance.
- The effort for deployments of the simulation system on different machines should be kept low. If the target of a tool comprises a large group of persons the distribution should be possible without a costly setup process.
- Support for common network protocols like SOAP[1], REST[2] and RPC[3] in order to set up a distributed application would be favorable.
- Support for matrix algebra and numerical operations should easily be available with the platform. Results are often kept in matrix form which requires simple access to particular rows and columns or subsets of the matrix. Also support for re-ordering operations is useful e.g. if results of multiple simulation runs are kept in a single matrix.
- A good documentation and a large community are helpful during the development process. Whenever a problem occurs with a particular function or an unexpected error is raised any support is welcome. With common tools many problems have been solved before by somebody else.
- License costs are usually not an issue at a university if academic licenses exist but they are important if a commercial exploitation is intended.

---

[1] Simple Object Access Protocol, the most common protocol for web service calls.

[2] Representational State Transfer, an HTTP-based protocol that allows the exchange of information between two machines in a computer network. It can be used for web service implementations.

[3] Remote Procedure Calls. predecessor of SOAP.

- Support for easy processing of input files and the manipulation of result files is desirable. In particular during the development phase the implementation of a mechanism for reading input files should be easy. Thus support for text file operations is helpful.

The following table gives a summary of how well three programming platforms, C++, Matlab and Python meet the requirements mentioned above:

Property	C++	Matlab	Python
Object-orientation	+++	+++ (since 2008)	+++
ODE solvers available	+ Bindings for many common solvers are available, but setup work in order to use them may be significant. Writing a proprietary solver can lead to good results.	+++ Many common solvers work out of the box and are easy to use.	+++ Libraries with many common solvers work well and are easy to use.
Performance	+++ Good code is very fast.	+ Solvers are precompiled and fast if no interaction with script-based code is needed.	+ Solvers are precompiled and fast if no interaction with script-based code is needed.

Multiple deployments	+++ Once the setup package is ready, deployments on other machines should be no problem.	- Only possible on machines with a Matlab runtime installation.	+ A python interpreter is needed. It can be delivered together with the code in setup packages for an uncomplicated installation procedure.
Support for SOAP, REST, RPC	- Poor. There are some implementations but getting them to work can be tricky.	-- No support for server implementations. Client implementations are possible.	+ Many third party libraries work well with Python.
Support from community and documentation	- Everybody knows C++ but its dissemination in the scientific community has rather decreased during the recent years. Support for some specific problems may be difficult to get.	++ Widely spread among universities and the scientific community. Help is never far.	+++ Excellent. Hardly any Python-specific problems that have not been solved yet.
License costs	+++ Free.	--- Expensive.	+++ Free.

Table 1: Comparison of C++, Matlab and Python in terms of qualification for building simulation

Due to the above considerations Python was chosen as a programming platform. Miller (2013) stresses the qualification of Python for building simulations not least because of its versatility and the number of available libraries that address many problems occurring in the course of development.

3.2 Transferring the model into code

One of the most important requirements for the implementation of RC models is a strict object-orientated mapping between zones and construction elements and nodes of the relevant RC model. The classes that were set up in the code are displayed in Figure 11.

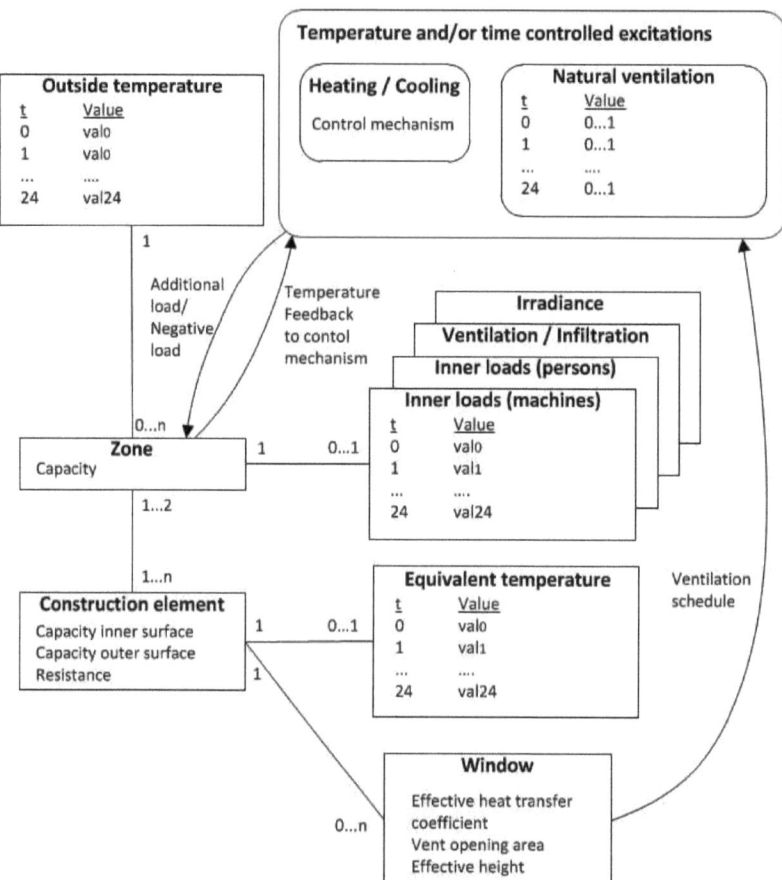

Figure 11: Simplified UML chart of the implementation of the RC model.

The "Zone" type of object represents one or more rooms of the building model. The most obvious approach would be to create one zone per room. A zone is specified by the value of the lumped heat capacity that is assigned to it. Also a number of internal load objects are attached to it. A zone can consist of an arbitrary number of construction elements and by a specified number of internal load objects.

Construction elements dispose of two lumped capacities and one thermal resistance between the two surfaces. A construction element is linked with at least one zone. If a construction element separates two zones a second zone is linked to one of the surfaces.

Windows are attached to their parent walls. They contribute to the radiative temperature (eq. 2.7) and are specified among others by the vent opening area and the effective height. These values are needed to calculate the airflow rate (eq. 2.2) caused by natural ventilation. A list of openness coefficients between 0 and 1 can be specified for each window in order to create a schedule for natural ventilation. E.g. during the day people may not be present in residential buildings. Thus it is unlikely that windows will be opened even if natural ventilation were favorable for achieving a good indoor climate.

Internal loads are linked to zones. They are made up of a list of key-value pairs, the key specifying the time of impact and the value the intensity of impact. For a periodic steady state calculation typically 24 key-value pairs are supplied, one per hour. For an annual simulation 8760 key-value pairs would typically represent a TMY file.

Values in between two points of time are either interpolated or the relevant key is rounded down. E.g. if a window is opened at 7 a.m. the openness coefficients between 6 a.m. and 7 a.m. would be the same as for 6 a.m. instead of being interpolated. By contrast values of the outside temperature would be interpolated between two full hours.

Instead of giving a list of key-value pairs it is also possible to specify a continuous time function for an internal load instead of key value pairs.

Some excitations depend on the difference of the temperature between the outside and the relevant zone temperature. An obvious heating control mechanism for instance will switch the heating on if the zone temperature drops below a certain value. A similar control mechanism would be applied to a mechanical cooling system to avoid overheating.

Table 2 holds a list of objects used throughout the implementation including their most important physical properties and the cardinality with respect to other objects.

Object type	Cardinality	Description
Zone	Zone : Outside temperatures = n : 1 Zone : Construction elements= 1 : n Zone : Internal load object = 1 : 1	Represents one or more rooms. An internal heat capacity representing room air and the furnishing is assigned to it. It holds information regarding heating and cooling strategies.
Construction element	Construction elements : Zone = n : 1 Construction element : Windows = 1 : n Construction element : equivalent temperature = 1 : 1	Construction elements can represent an exterior or a partition or an internal wall or the floor or the ceiling. They are assigned a capacity on each surface and a thermal resistance. Exterior walls are assigned a list of equivalent temperatures.
Window	Windows : Construction element = n : 1 Windows : openness coefficients = 1 : 1	Windows are also assigned an effective heat transfer coefficient and a list of openness coefficients. The effective heat transfer coefficient is used for the calculation of the surface temperatures according to 2.3.
Internal load object	Internal load object : Zone = 1 : 1	Internal load objects can dispose of an arbitrary number of key value pairs. Typically they would hold 24 or 8760 elements. Instead they can also implement a function returning the relevant value with respect to the point of time. Internal load objects include the types of loads specified in Figure 11.

Outside temperature	Temperature object : Zone = 1 : n	The outside temperature is the same for all zones. Like with internal load objects the outside temperature object would typically consist of a list of 24 or 8760 elements. It can be created from a simple .csv file or a TMY file.

Table 2: Specifics of the individual objects used in the implementation of the RC model

3.3 Preprocessing

Running a simulation requires that the loads forming the excitations of the system are known or can be calculated based on the relevant zone temperatures if a temperature-driven strategy is applied. Some of these loads can either be calculated at runtime or the relevant values are supplied before the start of the simulation. For instance solar gains depend on the meteorological conditions as well as on the surroundings that cast a shadow unto the relevant building. Since the intensity of irradiation is independent of the inside temperatures resulting from the actual simulation it could also be calculated before the simulation is even started. The same applies for most of the internal gains like energy emitted by people and machines and for the meteorological conditions that are modeled via the equivalent temperature according to chapter 2.5.

Calculating theses values prior to the simulation run allows a load transfer from the simulation server to another machine because there is no time coupling of the relevant equations. In some cases, e.g. in particular in case of shadow calculation, the distribution of calculations to different machines can cause a significant performance gain.

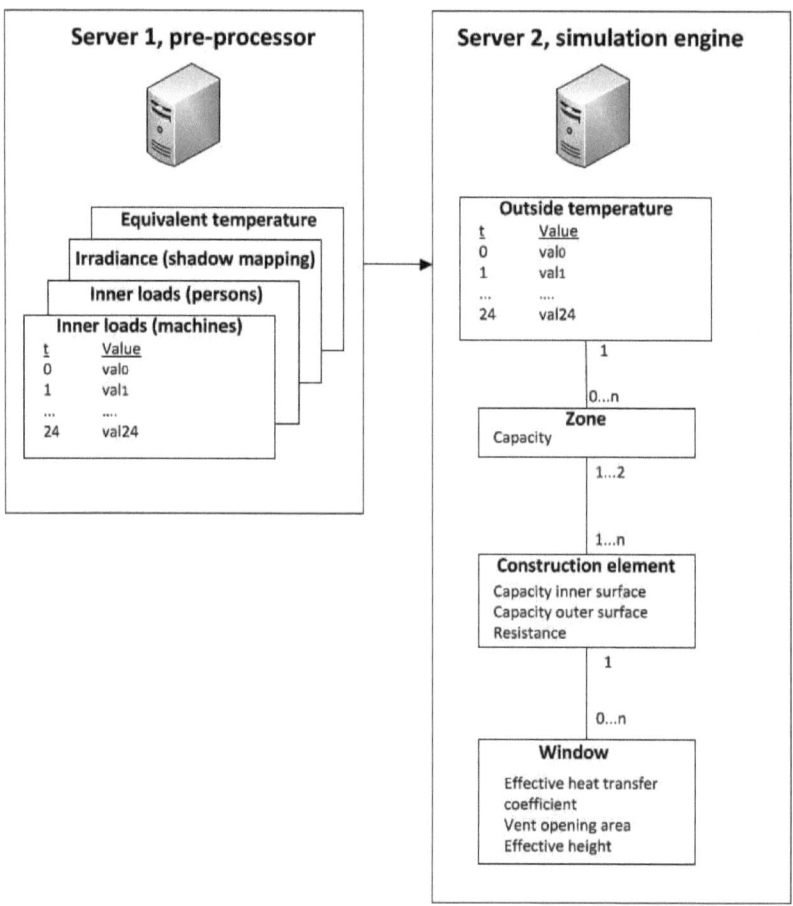

Figure 12: Schematic representation of processes distributed on different machines.

Figure 12 gives an overview of the parts that can be calculated independently of the actual simulation and that could be transferred to another machine. If one of the servers should represent a bottleneck more machines could be set up in order to distribute the calculation processes even further.

3.4 From CAD to RC models

The whole process from creating a building model to obtaining simulation results is displayed in Figure 13. Typically the first type of a building model will not be the RC model that is passed on to the solver for simulation purposes. It will rather be some type of CAD model that is set up with a program that is not designed to generate RC models. The approach to generate a RC model at the same time as the relevant CAD model is tempting and there have been efforts for an automated translation of CAD models into RC ones (Jones et al. 2013) but so far there have only been prototypes and no solution has made the breakthrough on the market. Some of the reasons therefore are:

- The creation of both types of models, CAD and RC ones is a complex matter. A tool aiming at creating both models at once runs the risk of spoiling the user interface due to different requirements.
- Setting up RC models requires a good knowledge in building physics which cannot always be expected from persons who are involved in the relevant CAD modeling.
- A separate setup of a RC model can be simple if CAD tools are capable of exporting geometry in an applicable format.
- Loosely coupled systems have also advantages regarding installation processes, complexity of software and attendance.
- User interfaces for building simulation need not be graphical ones. Models can also be neatly arranged if they only exist in text files.

The results can be returned either in the same form as the relevant RC model in order to have a simple mapping between the input and the output or in flat arrays. It depends on the further processing which type of format suits best.

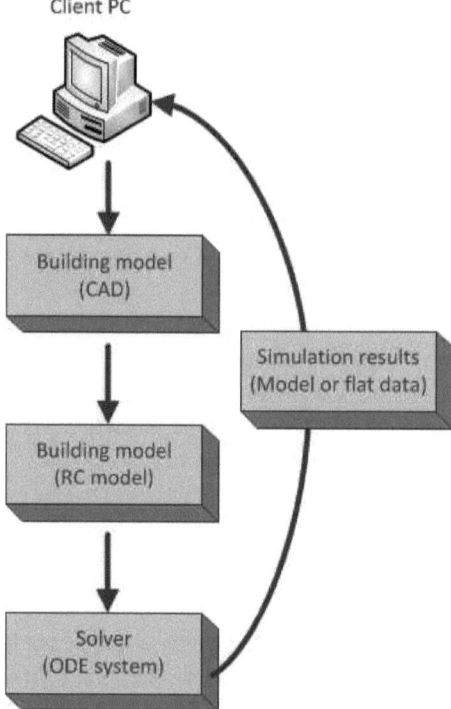

Figure 13: The original CAD model has to be transformed into a model that can be processed by the solver.

3.5 Heat capacity values

The number of research papers regarding suitable values of lumped capacity is significant. In many cases the setup of an RC model is tightly linked to the finding of suitable lumped capacity values.

Gouda et al. (2002) do not only focus on the analysis of model order reduction they also present an approach for calculating the value of the lumped capacity. First they calculate the capacities of a higher order model. These capacity values are then used with an optimization algorithm in order to find the capacities of a lower order model. Ramallo-González et al. (2013) calculate the resistances and capacities associated with the nodes of their model allowing for a dominant layer. Carlsaw (1950) provides an analytical solution for the value of the lumped capacity by assuming the profile of all excitations to have a sinusoidal form with a period length of one day. Since the calculation of a periodic steady state condition is a good starting point for further analysis and reduction algorithms of a multi-zone model, Carlsaw's (1950) approach is used for the calculation of the areal heat capacities in the following.

He chooses a Fourier transform for the temperature as a function of the time and the penetration depth. (See also Figure 14.)

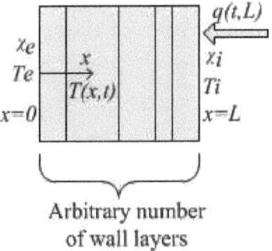

Figure 14: Carlsaw's approach is a function of time and penetration depth.

The calculation of the areal heat capacities is based on Fourier's law and the energy balance equations on both surfaces of the relevant construction element.

$$T(x,t) = \sum_{k=0}^{\infty} T_k(x) \cdot e^{ik\frac{t \cdot 2\pi}{t_p}} \qquad (3.1)$$

$T_k(x)$ denotes the amplitude of the temperature oscillation, t_p the period length. Thus on the inside surface of the wall the temperature reads

$$T_i = \sum_{k=0}^{\infty} T_{i,k} \cdot e^{ik\frac{t \cdot 2\pi}{t_p}} \qquad (3.2)$$

with $T_{i,k}$, the amplitude of the temperature oscillation inside and the wall temperature on the outer surface reads

$$T_e = \sum_{k=0}^{\infty} T_{e,k} \cdot e^{ik\frac{t \cdot 2\pi}{t_p}} \qquad (3.3)$$

with $T_{e,k}$ the amplitude of the temperature oscillation outside.

Applying Fourier's law,

$$c \cdot \rho \cdot \frac{\partial T(x,t)}{\partial t} = \lambda \cdot \frac{\partial^2 T}{\partial x^2}, \qquad (3.4)$$

the amplitude of heat flow, $q(x)$, at $x=0$ results to

$$h_i \cdot \left(T_{i,k} - T_{(x=0),k}\right) = -\lambda \cdot \frac{\partial T_k}{\partial x}\bigg|_{x=0} \qquad (3.5)$$

and the amplitude of heat flow $q(x)$ at $x=L$ results to:

$$h_e \cdot \left(T_{e,k} - T_{(x=L),k}\right) = \lambda \cdot \frac{\partial T_k}{\partial x}\bigg|_{x=L} \qquad (3.6)$$

The analytic solution for the temperatures T_i and T_e takes into account the construction layers of a specific construction element and allows the calculation of the area related energy:

$$Q = \int_0^{24h} \max\left(q(t), 0\right) dt \qquad (3.7)$$

Thus The areal heat capacity, χ_i, at the inner and the one at the outer surface, χ_e, can be calculated from:

$$Q = \chi_i \cdot \left(T_{i,k} - T_{(x=0),k}\right) = \chi_e \cdot \left(T_{(x=L),k} - T_{e,k}\right) \qquad (3.8)$$

A way to come to Carlsaw's (1950) results is given in EN ISO 13786. (CEN, 2007)

3.6 Solving the equation system

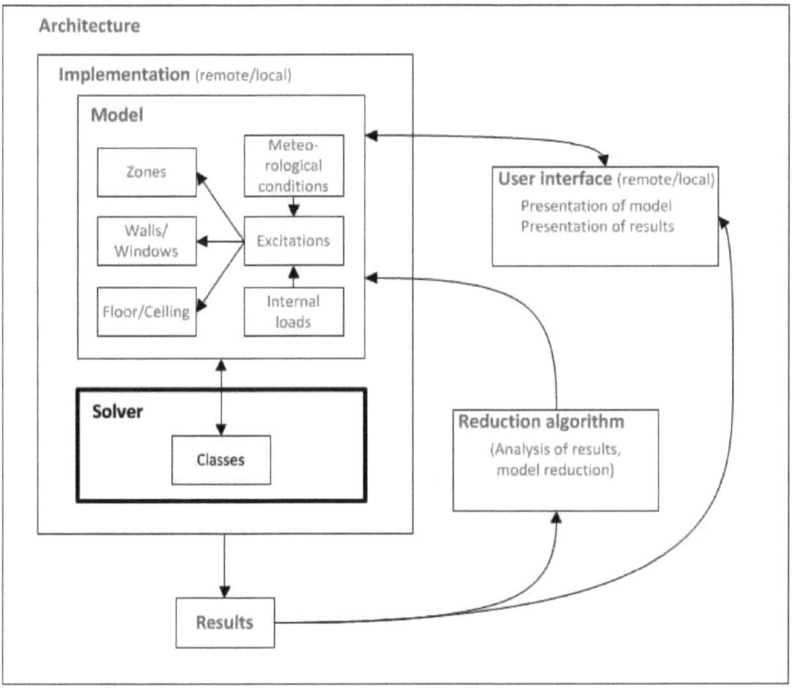

Figure 15: The solver represents the component that the model is passed for solving the relevant equation system.

Figure 15 displays the solver with respect to the whole implementation. It is not only passed the equation system resulting from the RC model representing the building but it also returns information relevant for the control systems, for instance heating and ventilation. The solver is vital for achieving a good performance.

3.6.1 State of the art in building simulation

Depending on the type of model an analytical solution is given (Peng and Wu, 2008; Nielsen, 2005) or a numerical solver is used to calculate the underlying differential equation system. (E.g. Goyal and Barooah, 2012)

The interest in finding the most applicable solver has been rather limited within the scientific community so far. Usually the choice of the solver remains a side note within research papers. This can be reasoned if the emphasis of the paper is rather put on the evolution of a new model than on tuning an existing one. Sometimes the solver only serves for a proof of concept and the details regarding a good performance are left aside or saved for later research. However in case of a RC model, that aims at raising the performance to an extent that annual simulations are possible in minutes at least, an efficient solver is vital.

The number of different solvers in science is quite high and a short overview of the classification shall be given in the following.

3.6.2 Step size

An important attribute of a solver regards the distance between points for which solutions are calculated. There are fixed-step solvers and variable-step solvers. Fixed-step solvers accept a particular step size and calculate solutions for the starting point plus an integral multiple of the step size.

Variable-step solvers adjust the step size depending on the requested level of accuracy. They use different time steps for different ranges of the equation system by setting them dynamically with respect to the characteristics of the system. For instance a frequent change of the excitations with respect to the system time may cause a reduction of the step size in order to meet the requirements for a certain accuracy of the solution.

Variable step-size solvers can have significant performance advantages and thus shorten the simulation time compared with fixed-step solvers because they extend the step-size in "smooth" parts of the simulation. In contrast the time-step for a fixed-step solver is set to a value in order to meet the requested level of accuracy at the most unfavorable time range of the simulation. Since by definition the step-size cannot be changed with fixed-step solvers the performance suffers because the same step size is also used in smooth sections.

Whereas with fixed-step solvers the time step has to be configured before the simulation is started it is the accuracy of the solution that is chosen with variable-step solvers. However the requested accuracy can only be maintained if for instance particular discrete excitations are taken into account by the solver.

From the scenario displayed in Figure 16 it can be seen that the excitation "Qk" is just between two points of evaluations. In this case the solution that is returned by the solver may not meet the required accuracy. This problem does not occur with fixed-step solvers if the time step is set to a value that enforces evaluations at all critical points.

Thus allowance for sudden consecutive changes in the intensity of impacts can easily be reached.

Another advantage of fixed-step solvers concerns the synchronization with real-time systems. Sometimes it is necessary to know exactly the points for which the solutions are generated. Setting the time-step to an integer factor of the required output interval is all that needs to be done in this case.

However also with variable-step solvers usually a limitation of the step size is possible as well as a configuration of "critical points" where a mandatory function evaluation should be performed.

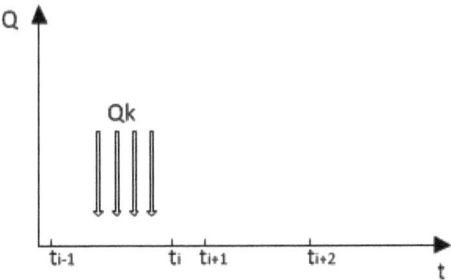

Figure 16: The load Q_k occurs between two points that the solution is calculated for. Thus it is not taken into account by the solver.

It was easily possible to reproduce the problem displayed in Figure 16 with a variable-step solver under certain constant temperature conditions. However with realistic models the solver never set the time-step to a value that ran the risk of skipping certain excitations.

For some reduction methods the periodic steady state of a model must be reached. (See chapter 4.4.) With the calculation of the periodic state also small differences in the step size between two simulation periods may lead to different results that would prevent the conversion to a periodic state. In this case the maximum step size must be limited.

3.6.3   One-step vs. multi-step

Solvers can be classified as one-step solvers or multi-step ones. One-step solvers calculate the solution at a certain point based only on the results of the calculation at the preceding point. Multi-step solvers base the solution at a certain point on the solutions and derivative values at several previous points with the goal of higher accuracy at an improved performance.

Runge-Kutta and Euler are examples for common one-step solvers whereas Adam's method belongs to the multi-step methods.

3.6.4 Implicit vs. explicit

Explicit solver algorithms base the calculation of the solution for a particular point only on the results and the derivatives of a previous state. Implicit methods calculate the state of a system at a future point and base calculations for the current time-step on both, current and the future states of the system.

Implicit methods often have an improved convergent behavior compared with explicit ones while explicit methods are less costly regarding the evaluations at a particular point.

An example for common explicit solvers is the Euler method, one for implicit solvers the improved Euler method. (Heun's method)

3.6.5 The choice of a solver

The choice of a solver is difficult because it is hard to assess the effectiveness of a particular solver in advance. The only way to evaluate solvers is to try them out and check the results for accuracy and measure the performance.

This can become tedious if several solvers shall be tried out and they cannot simply be plugged into the current implementation. Soetaert et al. (2012) give an overview of a variety of initial value problem solvers. They recommend using an implicit method if the problem is stiff and a predictor-corrector method or an explicit Runge-Kutta method if the problem is non-stiff. For problems which cannot be qualified as stiff or non-stiff they recommend e.g. the LSODA (2005) implementation which is based on Hindmarsh (1983) and Brown (1989).

The problems that building simulations bring along cannot be classified as stiff or non-stiff a priori because the intensity and frequency of excitations is not known in advance. Heating and cooling loads for instance may occur when the temperature drops below a certain value or exceeds a value respectively. Also the frequency of temperature dependent load changes cannot be clearly specified prior to a simulation run.

Python's SciPy comes with a variety of initial value solvers that can quite simply be used with the model at hand by setting up a differential equation system corresponding with the relevant model and passing it on to the solver.

Due to the discrete-time representation of many equation systems in research papers Euler's method is frequently used (Figure 17):

$$y_{i+1} = y_i + h \cdot y_i' \quad (3.9)$$

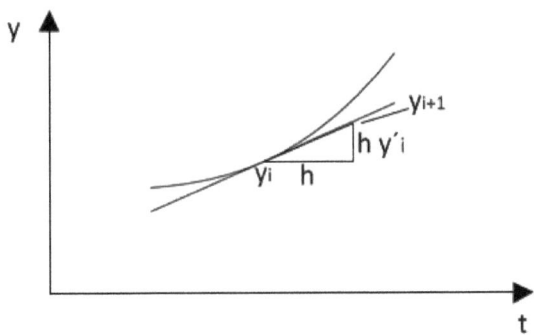

Figure 17: Euler's method extrapolates the deviation in order to calculate y_{i+1}.

Euler's method can easily be applied to equations 2.6-2.8 which represent a discrete-time form of the equation system. Whenever the choice of a time step is mentioned in research papers the chances are good that Euler's method has been applied. Aside from its wide dissemination Euler's method has also advantages regarding the integration with current implementations: No complicated third-party libraries have to be compiled and linked. Instead a self-written solver based on Euler's method is usually a matter of only a few lines in any programming language.

A modification of Euler's method is represented by the improved Euler method. (Heun's method) The tests that are carried out in the following were also made based on Heun's method. The results were similar in both cases.

3.6.6 Assessing different solvers

In the following several variable-step solvers and a fixed-step one were examined. Euler's method was chosen for the fixed-step solver implementation. In a first comparison LSODA (2005) was used to create a reference solution.

The model used to undergo the tests was set up according to 2.2. In fact it was kept simple in order to avoid unrepresentative effects on the performance of the solver. It was made up by a single rectangular-shaped zone bounded by six exterior walls with no internal loads resulting into a 7th order differential equation system. The set up was a "free-float" one with the building model exposed to a typical mid-European 24-hour-periodic winter day climate. Temperatures and solar gains between full hours are interpolated as well as equivalent wall temperatures. The details are left aside because they are of no relevance for the results.

3.6.6.1 Accuracy vs. step size

In the first step a 24-hour benchmark result for the zone temperature was generated with the tolerance of the solver set to 1.e-8. Then the Euler implementation was used with the same model and varying step sizes.

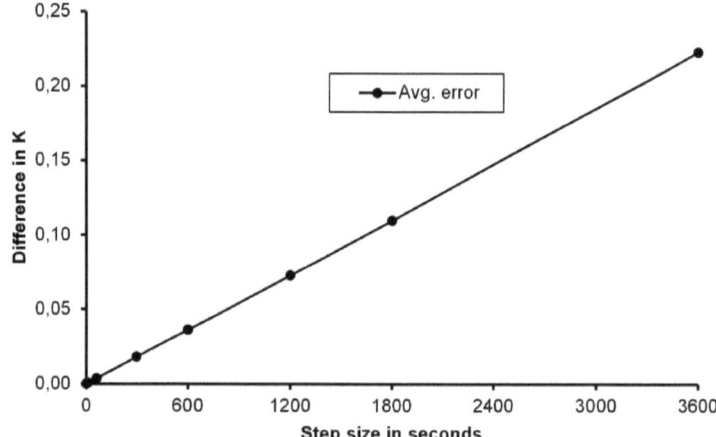

Figure 18: There is a linear dependency between the step size and the accuracy of results (represented by the average temperature deviation) with an implementation of the Euler method.

The accuracy is represented by the average deviation of the fixed time-step temperature from the benchmark solution mentioned earlier. The step size was extended in several simulation runs from 1 second to 3600 according to Figure 18. In fact step sizes in the order of 1800 seconds and above do not seem to make much sense in practice considering control systems rather operate within time-scales of minutes. But they can provide interesting information regarding the accuracy of a solver.

Similar considerations lead to a test run with the LSODA solver. While the step size has to be chosen explicitly with the Euler solver it is the accuracy that has to be specified with the LSODA solver. By contrast the step size is calculated automatically with a variable-step solver. Thus the axes of the chart in Figure 19 are swapped and the values for step-sizes are average ones.

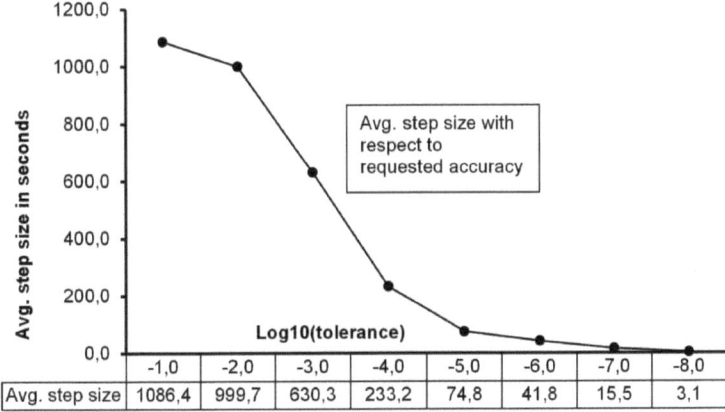

Figure 19: The linear correlation between accuracy and average step-size can only partly be observed with the LSODA implementation.

The linear correlation between the step-size and the accuracy of the temperature (which represents a direct outcome of the solution) can be observed in Figure 18. However the correlation between temperature and heating and cooling energy respectively is not necessarily linear. Thus another test was run where the temperature was held at a constant value and the amount of required heating and cooling energy was calculated. Again a benchmark simulation with the solver tolerance set to 1.e-8 was run. The deviations of the Euler-based results from the benchmark solution with respect to various time steps are displayed in Figure 20.

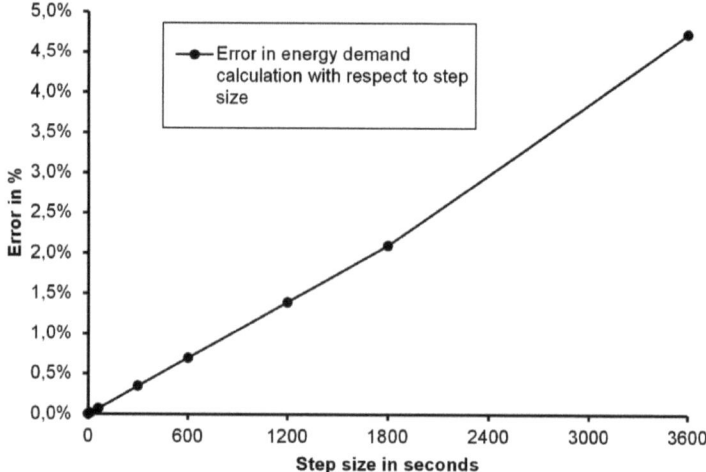

Figure 20: There is a linear correlation also between the accuracy of the energy calculations and the step size.

3.6.6.2 Performance evaluation

A universally valid evaluation of the performance of different solvers is difficult because the performance depends on many aspects of a particular model. A good performance of a solver with one model need not imply good performance with the next one. Nevertheless the calculation of the same free-floating model as above was run with different solvers and varying time steps.

Since the tolerance cannot be set explicitly with Euler-based solvers the values from Figure 21 are used as target tolerances for the LSODA and the BDF solver. The BDF solver is also part of a SciPy package and can easily be plugged into the existing implementation.

The test was set up in several steps:

1. The solver tolerance for the variable-step solvers (LSODA and BDF) was set to one of the values from Figure 18.
2. The time step for the Euler solver was set to the corresponding value.
3. The simulation period was set to the same value for all solvers. It was chosen in order to have simulation run times of several minutes at most.

4. Figure 21 displays the speed of the LSODA divided by the speed of the Euler solver and the BDF solver respectively.

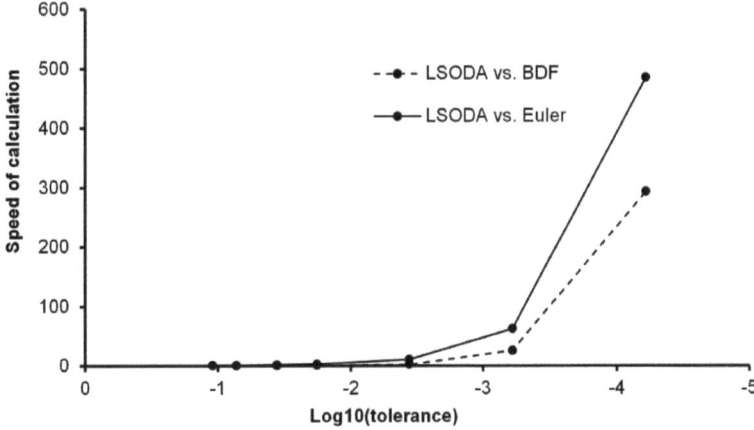

Figure 21: Speed of the LSODA solver compared with Euler solver and BDF solver respectively. Euler performs quite well as long as the required accuracy is low and time steps can be raised. LSODA outperforms Euler and BDF at almost any required accuracy.

3.6.7 Conclusion

Python's Scipy comes with a large set of initial value solvers that can be used with any linear differential equation system:

- LSODA (adaptive BDF or Adams solver)
- BDF
- Runge Kutta of different orders
- Adams

All of them can be configured with a maximum tolerance of result accuracy. Another advantage of these solvers is that they can be switched without any modification of the model that the equation system is based on. All that needs to be done is to change the parameters of the calling function. Thus several solvers were tried out with different scenarios in different situations and all of them worked reliably. The main criterion for the application of a particular solver was a greater performance than with others. However in most cases LSODA was applied as the default ODE solver since it partly takes over the task of choosing the right solver and it also proved to work reliably and fast.

One of the problems with Euler-based implementation concerns the uncertainty of the relationship between the size of the time-step and accuracy of results. Unlike with the solvers mentioned earlier it is not easily possible to choose a step-size with respect to a desired accuracy. There are several ways of addressing the problem:

- Georgios et al. (2013) treat the problem by reducing the step-size according to the change of temperature during a particular time-step.
- If a minimum accuracy of the results is desired tests can precede the actual simulation where the time-step is reduced until the relevant decimal place of the result does not change anymore. However this approach is expensive because it can multiply simulation times under unfavorable conditions. Additionally all critical simulation periods would have to be taken into account. However a priori these ranges are not always known.

3.7 Control systems

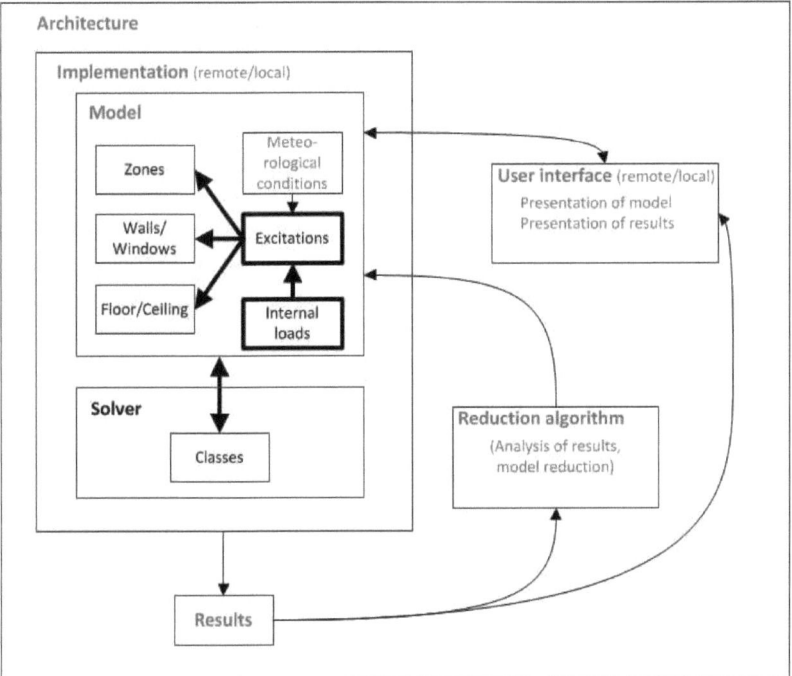

Figure 22: Control systems rather represent interactions between individual components than a particular component themselves.

Figure 22 represents the interactions between excitations, the RC model and the solver. The configuration of control systems is usually essential for loads whose intensity is not known prior to the simulation, such as heating and cooling and ventilation. They have to be deeply integrated with the implementation as a whole but they need to dispose of interfaces for configuration purposes, for instance temperature controlled heating and cooling loads.

3.7.1 State of the art, MPC

Heating and cooling systems as well as HVAC systems and natural ventilation can be controlled in several different ways. is The resident of a building who switches heating and cooling systems on and off and opens windows or keeps them closed involuntarily represents the most obvious thinkable control mechanism. However in buildings where control mechanisms have to be implemented in a large scale for many systems a completely subjective control strategy is not possible anymore.

The interest for model predictive control has grown during the last years. Camacho et al. (2004) were among the first to deal with control strategies that anticipate the behavior of the model due to excitations from outside. They coin the term model predictive control, MPC. Many researchers use RC models for MPC design. (Candanedo et al., 2013; Hazyuk et al. 2012, Lepore et al., 2013; Lehmann et al., 2013; Oldewurtel et al. 2012; Široký et al., 2011) Oldewurtel et al. (2012) focus on MPC based on weather forecasts. Široký et al. (2011) set up an MPC problem formulation which is constrained by comfort requirements and aims at the minimization of energy consumption. Lehmann et al. (2013) set up a model that also takes into account the air quality, room illuminance and TABS. Lepore et al. (2013) set up an MPC in that they minimize a cost function based on the primary energy heat flow rates and compare the results with those of a proportional-integral controller. They find out that a properly set up MPC algorithm can help save the total primary energy consumption.

The model presented in the thesis at hand aims at both, providing a platform for simulating completely user-controlled systems e.g. based on arbitrary timelines as well as trying out MPC strategies depending on the thermal behavior of a system. However cost function setup and optimizing is beyond the scope of the current research work.

3.7.2 Ventilation

3.7.2.1 Natural ventilation

Natural ventilation through windows is one of the most efficient means to control the temperature within a building. There is a significant difference in control strategies related to the time of the year: Whereas in summer natural ventilation is mainly applied to lower the room temperature it shall usually rather support short and fast air exchange in winter. If infiltration rates in a building are within the desired scope there should not be the need for open windows in winter. If residents decide to open them anyway it is likely because of a subjective need for fresh air or because of uncomfortable odors.

However in summer the thermal effect of natural ventilation through windows is significant. The airflow rate specified in eq. 2.1 represents the Buoyancy-driven ventilation. Wang and Chen (2013) use a mixed model that takes into account the Buoyancy effect as well as wind-driven effects and suggest several control strategies based on human behavior, indoor temperature and humidity compared with outdoor conditions. Neglecting wind-driven effects avoids overestimating cooling in summer because during warm summer periods no guarantees regarding nightly wind speed can be given.

Control strategies can be temperature-driven or time-driven or be a combination of both. Whereas in winter rather time-driven strategies are set up to support hygienic air circulation or to simulate rush airing, in summer rather temperature-driven strategies will be applied.

A uniquely temperature-controlled strategy during warm summer periods could be to open up the windows whenever the outside temperature drops below the inside temperature. However also during warm periods temperature-controlled strategies are likely to be constrained by certain schedules. For instance if residents are not present during the day a temperature-controlled strategy will not make sense if windows cannot be opened automatically. Thus to integrate the opportunity to configure schedules for temperature-controlled ventilation with a simulation tool is important.

In winter window opening periods will not be temperature-controlled but rather be kept as short as possible and probably be limited to rush airing. This behavior can be simulated by scheduling short daily periods of rush airing.

A third type of control strategy concerns buildings and climate conditions respectively where temperature variations are rather large or where solar irradiation is intense compared with outside temperature. This is where overheating is likely to occur but also where natural ventilation can cause a harsh drop of the room temperature. In this case a temperature-driven control strategy is implemented where windows open up if an upper bound inside temperature is exceeded and are closed if a lower bound temperature is undercut. (Figure 23)

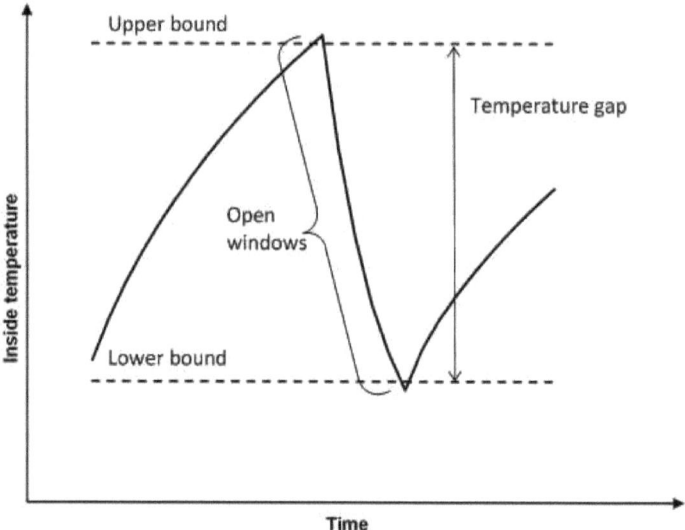

Figure 23: A temperature-controlled ventilation system needs to limit ventilation periods if the outside temperature causes a sharp decrease of the room temperature.

3.7.2.2 Mechanical ventilation

Control strategies for mechanical ventilation can be set up in a similar way as those for natural ventilation. In winter mechanical ventilation is rather turned on to support hygienic air exchange than to control temperature (unless the mechanical ventilation system disposes of a heat exchanger or incoming air is pre-heated). In summer a similar strategy as for natural ventilation can be set up.

A difference between natural and mechanical ventilation can be observed regarding the mode of operation from a technical point of view.

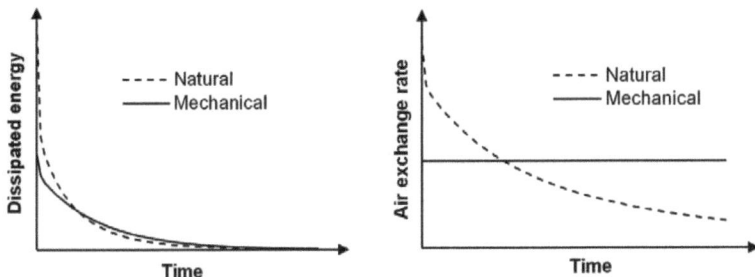

Figure 24: Schematic representation of typical dissipated energy and air exchange rates of natural and mechanical ventilation in summer.

Figure 24 displays typical curves for summer ventilation of a room where the inside temperature is above the outside temperature at the beginning and converges towards the outside temperature as the inside and the outside air intermingle. In the example the dissipated energy is higher with natural ventilation at the beginning because due to Buoyancy effect the air exchange is higher than with mechanical ventilation at the beginning. As the inside temperature falls the Buoyancy effect ceases and the exchange rate drops below the exchange rate with mechanical ventilation which is kept constant at all times.

3.7.3 Shading

Shading is an effective means to avoid summer overheating. Hviid et al. (2008) focus on both, the effects of shading devices on the internal daylight distribution as well as the integration of their shading model with thermal simulations. Nielson (2005) sets up a shading model for window recess and overhang. The German standard DIN EN 13363-1 (2007) contains methods of calculating the effects of outside and inside shading devices regarding the transmittance of irradiance. Applying external shading devices and special windowpanes the solar gains through glazing systems can be lowered by more than 80 %. It remains to note that shading devices usually do not have an effect on the solar energy absorbed by opaque elements. For the calculation of shadow mapping please refer to chapter 2.4.1.

Control strategies for shading devices aim at exploiting solar gains in cold seasons and at lowering them when indoor temperatures rise above a comfortable point. The integration of the relevant control strategy with the simulation at hand is achieved via the calculation of several time rows of solar gains prior to the actual simulation. Every time row represents the solar gains with respect to a particular configuration of the shading devices. This would include e.g. solar gains with blinds closed, solar gains with blinds opened, solar gains with blinds half opened... At simulation runtime the control mechanism selects the time row of solar gains with respect on the current inside temperature.

In the simulation at hand a single set point for switching between solar gains with and without shading devices was configured. Particular schedules can easily be taken into account by setting up the relevant time rows of solar gains.

3.7.4 Internal gains, heating and cooling

Internal gains comprise all gains from heat sources inside the building. Although solar gains are treated like internal gains once they have been calculated, they originate from outer excitations whereas internal gains are placed within the building.

The most common internal heat sources are people, machines and heating systems including the relevant installation devices. There are two major distinguishing characteristics with internal gains:

- The portion of convective compared with radiative input. Internal gains are split up into two parts. The first part represents the direct impact on the zone temperature and is summed up by Q_i in eq. 2.5. The second represents the impact on the surface temperature of construction elements and is summed up by $Q_{Rad,z}$ in eq. 2.7. For instance depending on the heating system radiators can be installed in a building or a convective heating. The former would result in a high radiative portion of the heating energy the latter in a high convective one. Another example that is relevant for different configurations concerns computers and notebooks compared with persons. Whereas computers and notebooks are taken into account with a high convective portion the energy emitted by persons is split up into two equal parts.
- The second characteristic concerns the relevance for control systems. Some internal gains are scheduled and are independent of the room temperature that is calculated in the course of the simulation, the intensity of others correlates with the room temperature in that it is controlled by the mechanism that is implemented. An example for the former are people in an office whose physical presence is usually independent of the thermal behavior of the building, the latter comprises the heating system that is adjusted with the profile of the inside temperature.

The Austrian standard ÖNORM B 8110-3 (2012) suggests values for internal heat sources like machines and persons.

The heat source that is less trivial for the implementation of a simulation system is the heating system. (All considerations regarding the control of the heating system are analogous for the cooling system.)

Several simulation strategies have been implemented. The most obvious is to switch on the heating system when the room temperature drops below a set point and to switch it back off when another (higher) set point is reached.

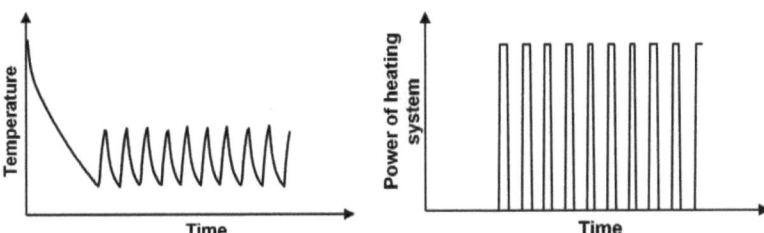

Figure 25: Typical temperature profile and the corresponding power profile of the heating for a model where a thermostat-controlled heating is simulated

On the left side of Figure 25 the temperature profile of a zone is displayed where the set point for the activation of the heating system is not identical with the point where a comfortable room temperature is reached. Between the two points the heating system is set to a power stage that is strong enough to raise the temperature of a room up to a comfortable point. Then it is switched off until the temperature drops below the set point again. On the right hand side there is the corresponding power profile of the heating. The smaller the gap between the two set points the more often the heating system is switched on and off.

If the temperature gap between the two set points is shrunk and the two points are finally merged to a single point the solver will become instable because the heating system would have to be switched an infinite number of times per time unit.

An improved control mechanism can be implemented if the power of the heating system is assumed to be infinitely variable. In this case the temperature is kept constant and the amount of required heating energy is continuously calculated.

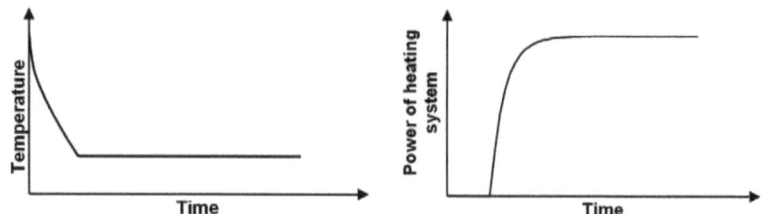

Figure 26: Typical temperature profile and the power profile of the heating for a model where the ideal heating power is continuously calculated

Figure 26 displays a typical temperature profile of a zone with an ideal heating system. For simulation purposes it is often the demand of heating energy and the maximum heating capacity that are of particular interest. The demand of heating energy is the same in both configurations. However the maximum heating capacity is unknown in the thermostat-controlled case. It has to be set to an estimated value. Only the resulting temperature profile reveals whether the estimated value was a good guess. If the temperature rises very fast the capacity of the heating system may have been overestimated. If it rises rather slowly or even decreases despite the heating system running at the max the capacity is estimated too low.

The ideal heating system displayed in Figure 26 returns the exact value of the heating capacity at all times. The implementation of an ideal heating system requires some conversions of the differential equations that the relevant model is based at. The approach is based on the idea to keep the relevant zone temperature constant and to calculate the heating or cooling energy. Only when the sign of the input energy swaps the zone temperature is released again.

The change of the zone temperature is represented analogously to eq. 2.6 and converted to result to 0:

$$C_Z \frac{\partial T_Z}{\partial t} = Q_{i,conv,Z}(t) + Q_{sol,conv,Z}(t) + Q_{vent,Z}(t) + \sum_j \frac{T_Z - T_{i,j} + \sum_k T_Z - T_{w,i,k}}{R_j} = 0 \quad (3.10)$$

$T_{i,j}$ denotes the temperature at the inner surface of construction element j and $T_{w,i,k}$ the temperature at the inner surface of window k.

$$Q_{i,conv,Z}(t) = Q_{pers,Z}(t) \cdot cf_{conv,pers} + Q_{mach,Z}(t) \cdot cf_{conv,mach} + Q_{heat,ideal,Z}(t) \cdot cf_{conv,heat} \quad (3.11)$$

$Q_{i,conv,Z}(t)$ represents the convective portion of the internal loads including potential heating and cooling loads as well as the convective portion of internal gains through persons, internal gains through machines. Accordingly $cf_{conv,pers}$, $cf_{conv,mach}$ and $cf_{conv,heat}$ denote the convective amount of energy by persons, energy by machines and energy by heating system respectively.

The equations relevant for representing the temperature at the inner surface of window k have already been set up in chapters 2.2 and 2.3 respectively:

$$T_{Rad,Z} = \frac{\sum_j A_j \cdot T_{i,j} + \sum_k A_k \cdot T_{w,i,k} + Q_{Rad,Z}(t)}{\sum_j A_j + \sum_k A_k} \qquad (2.7)$$

$Q_{Rad,Z}(t)$ represents the radiative portion of the internal loads analogously to eq. 3.11:

$$Q_{Rad,Z}(t) = Q_{pers,Z}(t) \cdot (1 - cf_{conv,pers}) + Q_{mach,Z}(t) \cdot (1 - cf_{conv,mach}) + Q_{heat,ideal,Z}(t) \cdot (1 - cf_{conv,heat})$$
(3.12)

$$T_{w,i} = T_{eff,i} - \frac{U_{eff}}{h_{eff,i}} \cdot (T_{eff,i} - T_{eff,e}) \qquad (2.30)$$

$$T_{eff,i} = T_Z + \frac{h_{r,i}}{h_{eff,i}} \cdot (T_{Rad,Z} - T_Z) \qquad (2.21)$$

In accordance with equations 3.10-3.12, 2.7, 2.21 and 2.38 $Q_{heat,ideal,Z}$ can be calculated. The conversions have been made by feeding the equations into the math program Maxima. First an explicit representation of $T_{Rad,z}$ has been calculated and inserted to get an explicit expression of $Q_{heat,ideal,Z}$.

The explicit representation of $Q_{heat,ideal,Z}$ has another advantage regarding the performance of the solver. The average step size of the solver with the explicit representation of $Q_{heat,ideal,Z}$ was up to ten times as high as with the thermostat-based configuration of the heating and cooling system which results in shorter calculation times.

Alexander et al. (2008) have designed a set of test cases suitable for testing simulation tools. MZ320 is used to demonstrate the functionality of the ideal heating and cooling system with a three-zone model. The specifics are displayed in Figure 27 and in Table 3 respectively.

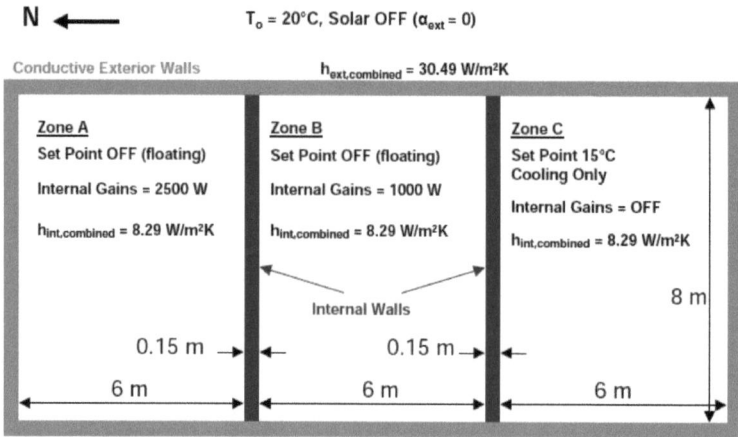

Figure 27: Dimensions of the model used with BESTEST MZ320 (Alexander et al., 2008)

Element	Conductivity (W/(m·K))	Thickness (m)	Conductance (W/(m²·K))	Resistance (m²·K/W)	Density (kg/m³)	Specific Heat (J/(kg·K))
Int Combined Surf Coef			8.2900	0.1206		
Common Wall Material	1.20	0.15	8.0000	0.1250	1400	1000
Int Combined Surf Coef			8.2900	0.1206		
Total air-air			2.7303	0.3663		

Table 3: Material properties of the walls used with BESTEST MZ320 (Alexander et al., 2008)

The test case allows for the calculation of the analytic solution which facilitates the verification of the simulation results with the correct solution.

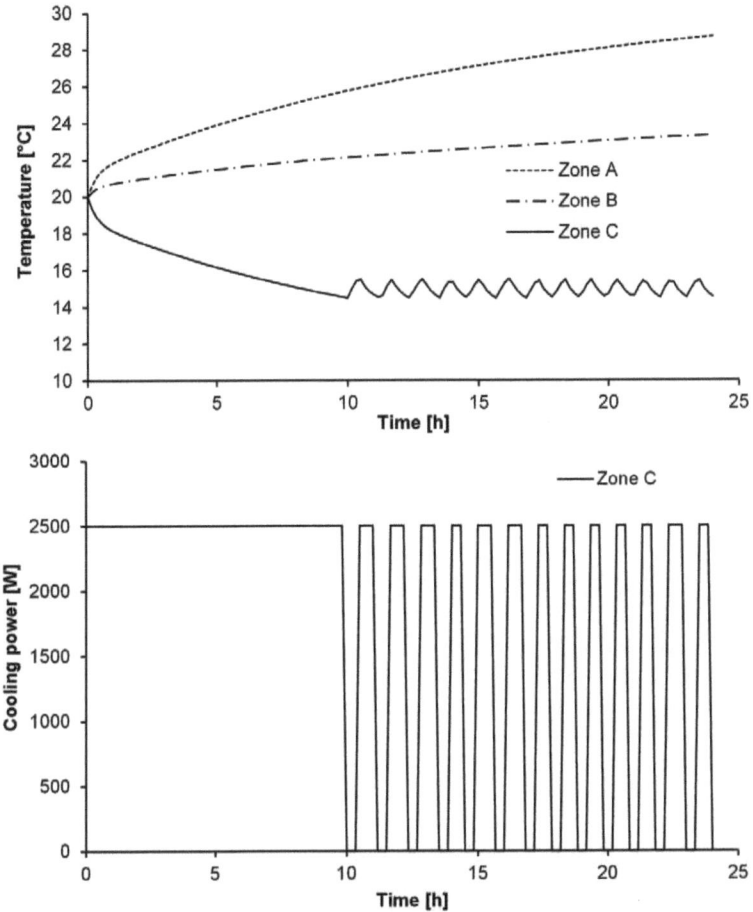

Figure 28: Temperature profile and relevant cooling power profile of zone C according to BESTEST MZ320 (Alexander et al., 2008) in case of thermostat-based control

Figure 28 displays the temperature profile of all three zones. A test condition for zone C reads that the cooling set point is at 15 K. The other two zones are configured as free-floating ones. The upper part of Figure 28 displays the temperature profiles converging to the steady state temperature conditions. The lower part displays the amount of withdrawn energy. The system is configured to extract energy with a capacity of 2500 W until a temperature of 14.5 K is reached. Then it is switched off and switched back on only when the temperature has risen to 15.5 K. The exact cooling power that would be needed in a steady state condition is unknown a priori with the thermostat-based configuration. It could only be iteratively determined in the course of the simulation.

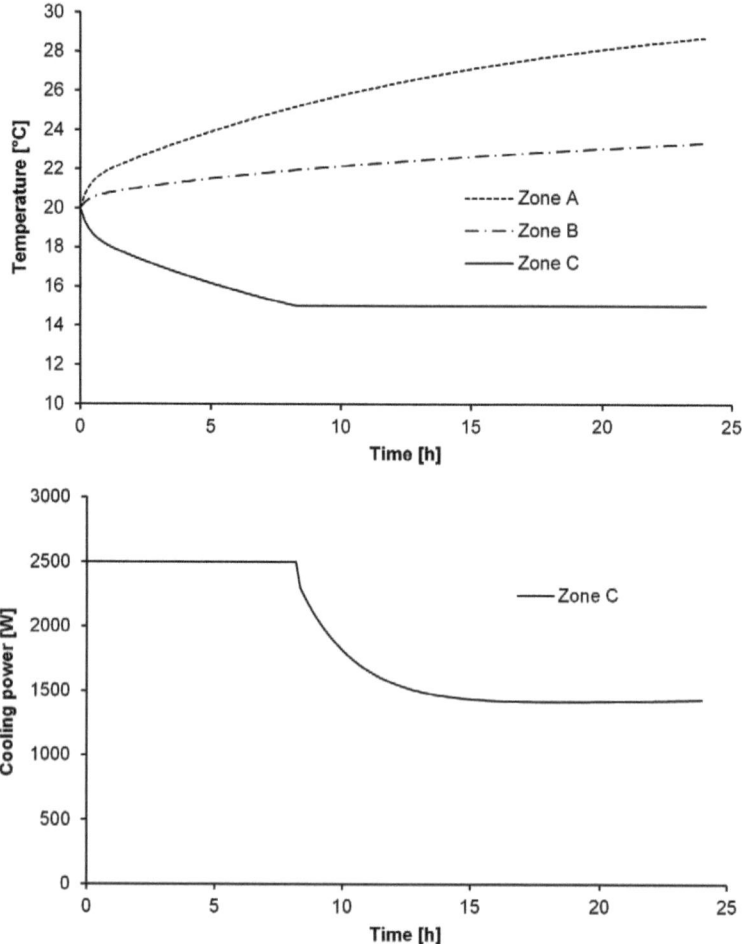

Figure 29: Temperature profile and relevant cooling power profile of zone C according to BESTEST MZ320 (Alexander et al., 2008) in case of an ideal cooling configuration

The test scenario is the same in the next case but the configuration of the cooling system of zone C is different. The system is configured to withdraw energy with a capacity of 2500 W until a temperature of 15 K is reached. The cooling system is configured to operate in ideal mode then and to withdraw just as much energy as to keep the temperature constant. If the temperature in zone C dropped below 15 K due to the interactions with the neighboring zones or due to meteorological excitations from outside the ideal cooling system would be switched off.

It is also possible to configure an upper bound for the ideal heating to become effective and a lower bound for the ideal cooling. That way the minimum heating and cooling power can be calculated in a simulation run as well as the exact demand of heating and cooling energy.

3.7.5 Conclusion

Control systems are believed to offer quite a potential regarding the total energy consumption of a building. Recent research papers present control strategies that even allow for weather forecasts in order to coordinate all control systems. RC models are best suited for research in this field because they remain manageable also if many sources of impact are modeled in the context of a large building model. They can easily be adapted if different control strategies shall be tried out and simulations can be run in an acceptable amount of time if model reduction is applied in a sensible way.

Different control strategies have been investigated in this chapter. Time-driven control strategies are easier to implement because they do not allow for the current zone temperature. In contrast to temperature-driven strategies this can be advantageous regarding the overall performance because the excitations do not have to be adjusted at runtime. However realistic scenarios demand for temperature-driven control strategies or a combination of both, temperature- and control-driven strategies.

System	Temperature-driven?	Time-driven?	Remark
Ventilation	Yes, with respect to given schedules in summer.	Yes, with respect to room temperature in winter.	In summer a temperature-driven strategy is more relevant, in winter a time-driven one.
Shading	Yes.	Rather not. (Solar	Either manual or

		gains or glare can require additional algorithms.)	automatic operation of blinds, etc.
Internal gains (people, machines)	No.	Yes.	Presence of people and computers etc. is usually independent of room temperature.
Heating and Cooling	Yes.	Not necessarily.	Temperature-driven control strategies are more relevant.

Table 4: Summary of different control strategies.

Table 4 contains a summary of control strategies with respect to different systems. Ventilation control needs to be implemented allowing for instabilities of the solver that can occur if ventilation causes a harsh drop of the room temperature.

Two control strategies for heating and cooling systems have been outlined. If the second one is applied the exact demand for heating and cooling energy can be calculated as well as the required capacities. The application of the ideal control strategy also has advantages regarding the overall performance of the implementation.

4 Reduction algorithm

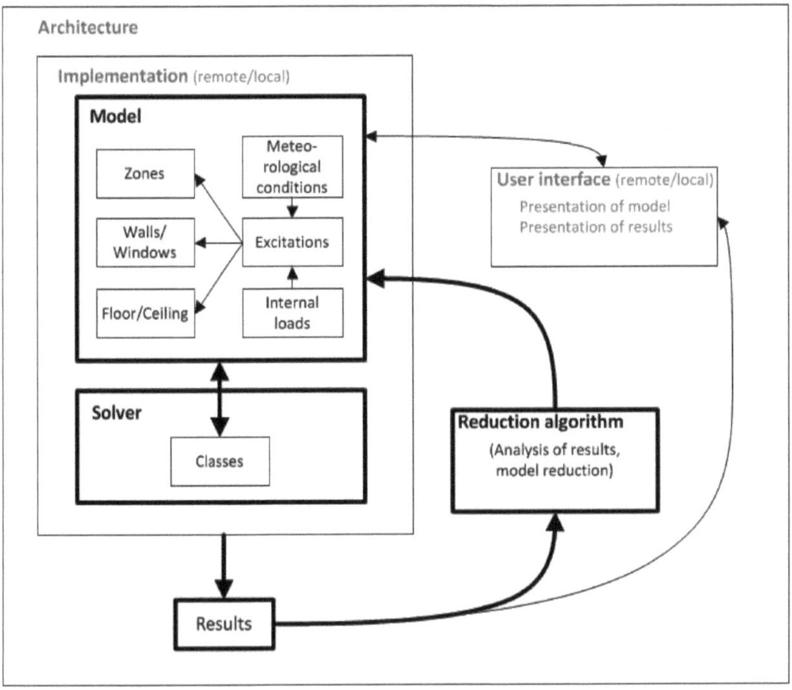

Figure 30: The reduction algorithm can remain independent of the implementation of the model and the solver.

Figure 30 displays the implementation of a reduction algorithm with respect to other parts of the implementation. It is located between at least one pre-simulation and the actual simulation. The outcome of the reduction algorithm is relevant for the setup of a new reduced model which may either be pre-processed once again or undergo the simulation that is intended by the user.

4.1 State of the art in performance tuning

The performance of a simulation depends on several characteristics of the equation system, for instance the rate of sudden changes of the intensities of excitations or the impact of temperature-controlled excitations. Also the requested tolerance of result accuracy contributes to a good or less favorable performance. However there is usually little scope in the adaptation of these parameters.

An important solver-specific configuration detail that contributes significantly to the simulation performance and the accuracy of results is the time-step that the solver is set to. While variable time-step solvers calculate the size of the time-step automatically it can be set explicitly with fixed step solvers. Dos Santos and Mendes (2004) do not only model thermal behavior but also moisture transfer. They experiment with different time steps in order to reach a better performance.

Another characteristic that plays an important role regarding the performance of the simulation is the order of the equation system. According to Kramer et al. (2012) there has been a lot of research concerning model order reduction. They observe that some approaches are based on the reduction of full-scale models and others start with a simplified model from scratch. Also Gouda et al. (2002) aim at the reduction of the model order. They investigate the number of nodes per construction element that a lumped capacity is assigned to and apply an optimization method to identify capacities and resistances of a smaller order model in order to maintain the accuracy of results. They end up with a second order model with two capacitances and three resistances (*3R2C*) per construction element.

Even earlier Hudson and Underwood (1999) assign a lumped capacity to every construction element. Antoulas et al. (2001) present a comparison of seven model reduction algorithms in their survey. All of these algorithms depend either on singular value decomposition or on moment matching based methods. Fraisse et al. (2002) achieve good results by aggregating multiple walls into one *3R4C* model. Goyal and Barooah (2012) set up a non-linear model because of taking into account the effects of moist air on temperature. Thus the order that they end up with in the full-scale version is higher than with modeling only heat transfer. They propose a reduction method that exploits the fact that the non-linear terms of their system only affect a small number of states. In their reduction method they make use of a balanced truncation to reduce the order of the model. In their tests they focus on rather short simulation periods than on months or even years.

What many research papers have in common is that reduction methods only start after the model has been set up. Hardly any researchers investigate how a reduced model can be set up with respect to the geometric and physical properties of the relevant CAD or 3D building model. However this type of optimization has a great potential because engineers can intuitively tag certain parts of the model as less important and more likely to be reduced in the model. Additionally an appropriate algorithm can identify regions of a building where detailed modeling leads to longer simulation times but to no significant gains in result accuracy.

Thus the approach of reducing the order of a model in order to achieve significant performance gains is pursued also in the current chapter of this thesis because it is a means to reduce simulation times independently of the underlying solver or conversion algorithms. At the same time geometric and physical circumstances shall be exploited the best possible way before the model is set up.

4.2 Introduction

The model that has been described in chapter 2 is used as a basis for the upcoming performance tuning algorithms. In the simplest thinkable form every construction element produces two equations and every zone one. Chapters 2.2.2 and 2.2.3 contain instructions about how to save equations with internal walls and partition walls under periodic conditions.

These optimizations do not cause a loss of accuracy compared with the full-scale model. They can even be fully automated and lead to a significant performance gain in particular if only some zones of a building are modeled and the rest of the building is approximated via the assumption of periodically repeated zones.

Chapter 4.3 contains another reduction method that can be applied to full-scale models as well as to reduced-order ones. Also in this case there is no loss of accuracy.

The reduction methods outlined in chapters 4.4 and 4.7 are less trivial. They are intended for different simulation goals. In one case the demand of heating and cooling energy is calculated with a simulation run in the other case the temperature profile of particular zone is needed. In both cases there is a tradeoff between performance and accuracy. The results from simulations of several test scenarios will be analyzed in order to allow a quantitative assessment of the loss of accuracy.

The reduction methods outlined in the following differ not only regarding the loss of accuracy but also the potential of being applied repeatedly to the same model. An idempotent method can be applied multiple times and does not modify the result beyond the initial application. The reduction method presented in chapter 4.4 can lead to an (unwanted) additional model reduction if it is applied more than once.

Description	Order reduction	Chapter	Accuracy preserved?	Idempotent?

Internal walls can be modeled with a single equation (instead of two like asymmetric construction elements). Thus one equation per wall can be removed from the equation system.	n (n: number of internal walls)	2.2.2	Yes	Yes, can lead to further model reduction after zone and/or wall merging has/have been performed.
Internal walls between periodically repeated zones have the same temperatures at both surfaces. Thus one equation per wall can be removed from the equation system.	n (n: number of internal walls)	2.2.3	Yes	Yes, but effective only once.
Wall merging can be accomplished with walls that have identical physical properties and are adjacent to the same zones.	2n (n: number of eliminated walls)	4.3	Yes	Yes, can lead to further model reduction after each modification of the model.
Zone merging aiming at energy calculation. Can be accomplished with zones that have similar properties.	n (n: number of merged zones)	4.6	No	No, multiple calls can lead to "over-simplification".
Zone merging aiming at temperature profile calculation of a particular zone. Can be accomplished with zones that have similar properties.	n (n: number of merged zones)	4.7	No	Yes, but effective only once.

Table 5: Summary of reduction methods

Table 5 contains a summary of all reduction methods presented in the thesis at hand.

4.3 Wall merging-based model reduction

The idea of wall merging comes from the interpretation of individual walls as fragments of an original (parent) wall. The goal of the reduction method is to identify the fragments that can be merged to the (virtually) original wall that they are derived from. Wall merging includes basically two fragments per merge process. But repeated merging can lead to an association of several walls to a single new one.

RC models are based on the effective thermal resistance, *R*, and the effective heat capacity, *C*. The thermal resistance is equivalent to the inverse conductance, *K*, of the relevant construction element, thus a different representation of the thermal resistance reads: $R = 1/K$. Since *RC* models only take into account one-dimensional heat flow, *K* is in proportion to the surface area of the relevant construction element.

$$(T_i - T_e)/R = \sum_j (T_i - T_e) \cdot K \cdot fw_j \quad (4.1)$$

$$\sum_j fw_j = 1 \quad (4.2)$$

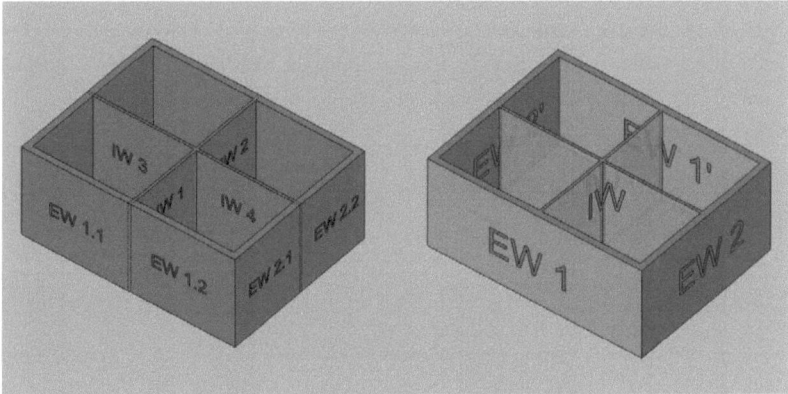

Figure 31: If walls can be merged because the relevant physical properties and excitations are identical equations can be saved.

Figure 31 displays a single zone model consisting of four rooms with partition walls in between. Modeling each wall separately would result into two equations per internal wall (or one per wall if symmetry properties can be exploited) and into four equations for each exterior wall.

The check whether walls can be merged is done in several steps:
- The zones located on both sides of the wall must be identical in both cases.
- The areal capacity must be identical at both surfaces.
- The areal conductance must be identical.
- In case of exterior walls the equivalent temperature according to 2.5 must be identical.

4.3.1 Wall merging

If all of these conditions are met walls can be merged by eliminating one of the two walls and executing the following steps:
- The total capacity of both walls is made up by the sum of the effective capacity at both surfaces the remaining wall.
- The total conductance is made up by the sum of the conductance of both walls.
- The area is made up by the sum of the area of both walls.
- In case of exterior walls all windows have to be attached to the remaining wall.

Applying wall merged reduction the inner walls can be represented by a single equation and exterior walls by two equations each. Assuming the external equations are identical from all four directions two equations could represent the complete exterior wall.

Status of model	Order	Details of order
Full-scale model	25	Zone: 1, exterior walls: 4x2x2=16, partition walls: 4x2=8
Symmetry of partition walls	21	Zone: 1, exterior walls: 4x2x2=16, partition walls: 4
Wall merging	10	Zone: 1, exterior walls: 4x2=8, partition walls: 1
Identical excitations from all directions	4	Zone: 1, exterior walls: 1x2=2, partition walls: 1

Table 6: Example for wall merging according to Figure 31

It would also be possible to accomplish wall merging if the difference between properties of walls is small instead of zero. The decision of whether the difference of properties is small enough can be based on two test simulations that are run prior to the actual simulation, one with the properties of the first wall and the other one with those of the second wall. The results allow to draw conclusions about the accuracy of the expected result.

4.4 Calculating the periodic state

The assessment of multiple zones regarding the similarity of their thermal behavior is relevant for the decision about which zones should be merged and which should remain unmerged. Aggregating the internal capacities of two existing zones in one and attaching all walls to it has little effect on the overall result if the thermal behavior of the original zones is similar. If the assessment is based on the comparison of the temperature profiles of two zones any distinct excitations including distinct inner loads need to be avoided. Also different set points for ventilation, heating and cooling will falsify the result.

Another impact on the resulting temperature profiles comes from the initial temperatures. Dos Santos and Mendes (2004) apply a "pre-simulation" (warm-up) to their model in order to reduce the initial condition influences.

The Austrian standard, ÖNORM B 8110-3 (2012), suggests to base the proof for protection against summer overheating on the temperature profile associated with the state of the model resulting from the daily periodic repetition of the outside excitations. The idea is adopted and the model is exposed to a repetitive 24-hour periodical combination of excitations by typical meteorological conditions, i.e. temperature and irradiation, until no more change in the temperature profiles of the individual zones at an interval of 24 hours can be observed. The condition for the steady state reads for a particular zone:

$$\max\big(\text{abs}\big(T(h)-T(h-24)\big)\big) < prec \cdot 2 \quad (4.3)$$

T(h) is the vector of air temperatures of all zones at the time *h*. Thus *T(h-24)* denotes the vector of air temperatures of the day before:

$$T(h) = \big[T_{z1}(h), T_{z2}(h), \ldots, T_{zn}(h)\big] \quad (4.4)$$

prec stands for the tolerance that has been configured with the solver. Once the steady state is reached the model is ready for applying a reduction algorithm.

4.5 Zone merging

Zone merging is an essential means to run the reduction algorithms which will be outlined in chapters 4.6 and 4.7. It can be achieved in two ways: By eliminating one zone and moving all relevant loads and walls to the zone that should persist or by eliminating both zones and creating a new zone where all walls and excitations from the original zones are placed. Under some conditions zone merging should not be accomplished because different excitations prevent a similar thermal behavior:

- The set points for cooling are different in both zones.
- The set points for heating are different in both zones.
- The set points for natural or mechanical ventilation are different in both zones.
- Schedules for natural or mechanical ventilation are different in both zones.

When two zones are merged the resulting one needs to undergo the following changes:

- All excitations of the two original zones need to be summed up and attached to the resulting zone.
- All walls including the windows are attached to the resulting zone.
- Internal heat capacities of both original zones are summed up and allocated to the resulting zone.

4.6 Model reduction for energy calculations

An algorithm is proposed that aims at the creation of a reduced RC model appropriate for calculating the demand of heating and cooling energy and of heating and cooling load. Thus the reduction algorithm rather focuses on the overall energy consumption than on the temperature profile of particular zones. The starting point for the algorithm is the periodic steady state that has been calculated according to 4.4. Then groups of zones with similar temperature profiles are formed. The condition for the similarity of the profile that has to be met at all times of the 24 h periodic state reads:

$$\max\left(\text{abs}\left(T_i(h) - T_j(h)\right)\right) < prec \cdot tol \qquad (4.5)$$

An arbitrary zone is chosen as first zone of a new group. Its temperature at the time h, $T_i(h)$, is compared with the one of the neighboring zones, $T_j(h)$. If the difference does not exceed a value proportional to the tolerance of the solver, zone j becomes a member of the new group.

The check is repeated for the next level of neighboring zones until no more zones meet condition 4.5. Then the next zone group is started with any zone that is not yet a member of a particular group.

prec denotes the absolute tolerance that the solver is set to. While the temperature profiles of two zones adjacent to each other cannot be expected to match within the range of the tolerance of the solver, *tol* has to be set to a value, bigger than one that controls the granularity of zone grouping. The bigger the value is chosen the more zones will be lumped together in a particular group. Trying out the reduction algorithm revealed that also rather big values of *tol* (100 and above) returned reasonable zone groups.

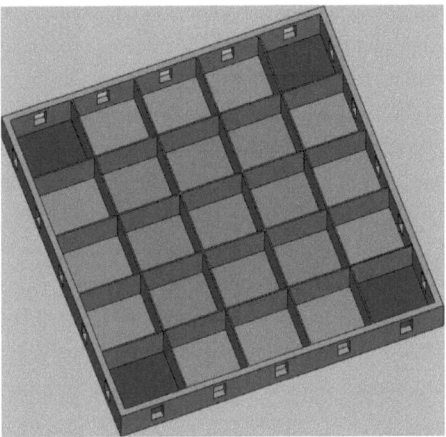

Figure 32: Zones are divided into groups. Zones are merged if their temperature profiles are similar.

Figure 32 displays an example consisting of 25 zones that the model reduction algorithm has been applied to. Zones belonging to the same group have been colored equally. The excitations from meteorological conditions in the above example have been configured to be identical for all four sides.

A repeated application of the proposed algorithm can cause an over-simplification of the model if the difference between temperature profiles of merged zones is small enough to cause further zone merging. In this case the reduced model could end up in a single zone model and the accuracy of results may suffer.

4.7 Model reduction for zone temperature calculations

Model reduction aiming at maintaining the temperature profile of a particular zone compared with the relevant full-scale model works different. Again the starting point for the algorithm is the periodic steady state that has been calculated according to 4.4. An iterative approach is followed for the identification of zones to be merged: the goal is to keep the temperature profile of a particular zone as close as possible to the profile of the same zone within the full-scale model. In the first step all zones are merged except the one whose temperature profile is needed. Then the periodic steady state is calculated again. If the temperature profiles of the full-scale model and the reduced one are close enough the model reduction is finished and the reduced model is ready for simulation. If the difference between the temperature profiles is too big, the zones adjacent to the zone of interest are also left unmerged. Figure 33 displays a schematic presentation of the progress of the algorithm: The zone labeled "1" represents the zone whose temperature profile is needed. In the first step all other zones are merged and the periodic state is calculated. Then the temperature profiles of zone "1" are compared with each other. In the example of Figure 33 they are not close enough yet, thus also the zones labeled "2" are left unmerged and the periodic state is calculated again. Again the temperature profiles of zone "1" are not close enough in the example at hand, thus also the zones labeled "3" are left unmerged and the periodic steady state is calculated. The temperature profiles of zone "1" are close enough in the third step of the example at hand. All zones labeled "1", "2" or "3" have to remain unmerged. So the identification of the most reduced model suitable for calculating a specific temperature profile requires a periodic steady state calculation for all preceding models. This can cause augmented calculation effort prior to the simulation run but can enhance the performance of a subsequent long term simulation.

The condition for the temperature profiles of a zone within a model at different reduction levels reads:

$$\max\left(\text{abs}\left(T(h) - T'(h)\right)\right) < prec \cdot tol \qquad (4.6)$$

T' represents the temperature profile of the zone as part of a reduced model and T the temperature profile of the same zone as part of the full-scale model. *prec* and *tol* have the same meaning as with model reduction according to 4.6. *tol* returned reasonable reduction results when it was set to a value between 2 and 10.

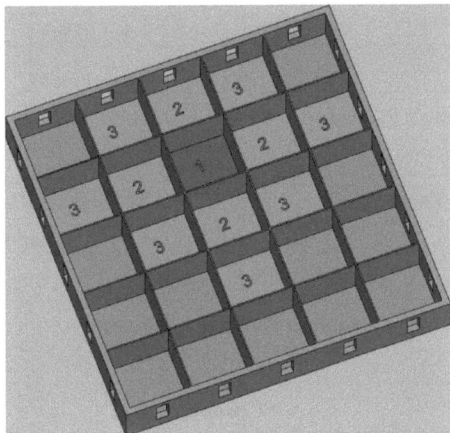

Figure 33: Neighboring zones are left unmerged in order to achieve similar temperature profiles with the full-scale model and the reduced one.

4.8 Results

Figure 34 displays a typical office floor that was used for evaluation purposes. Office rooms are located next to a corridor around the building and there is an inner courtyard. The floor is assumed to be repeated above and below an unlimited number of times. The specifics of the walls are listed in Table 7, those of the windows in Table 9. The internal heat capacity is identical for all zones. (Table 8)

Office rooms to the left and to the right are 4 m long and 3 m wide, the dimensions of the two large office rooms in the back are 4x5 m, the corridor has a constant width of 2 m. The whole floor is 3.5 m high.

The size of the windows on the long sides is 2x2 m. The dimension of those on the broad sides is 1.5x0.5 m and 1.5x0.75 m respectively.

Wall	Areal conductance (W K^{-1} m^2)	Areal heat capacity, inner surface (J m^{-2} K^{-1})	Areal heat capacity, outer surface (J m^{-2} K^{-1})
Exterior wall	0.190	16.607e3	56.188e3
Interior wall	3.077	75.236e3	75.236e3

Ceiling	0.738	325.955e3	93.754e3
Floor	0.738	93.754e3	325.955e3

Table 7: Particulars of the individual construction elements

Zone	Internal areal heat capacity $(J\ m^2\ K^{-1})$
Any	38.942e3

Table 8: Particulars of the zones

Window	U-value $(W\ K^{-1}\ m^{-2})$	g-value (-)
Any	1.185	0.62

Table 9: Particulars of the windows

Excitations by meteorological conditions were calculated from a TMY-file available for New York.

A graphical representation of the model is displayed in Figure 34.

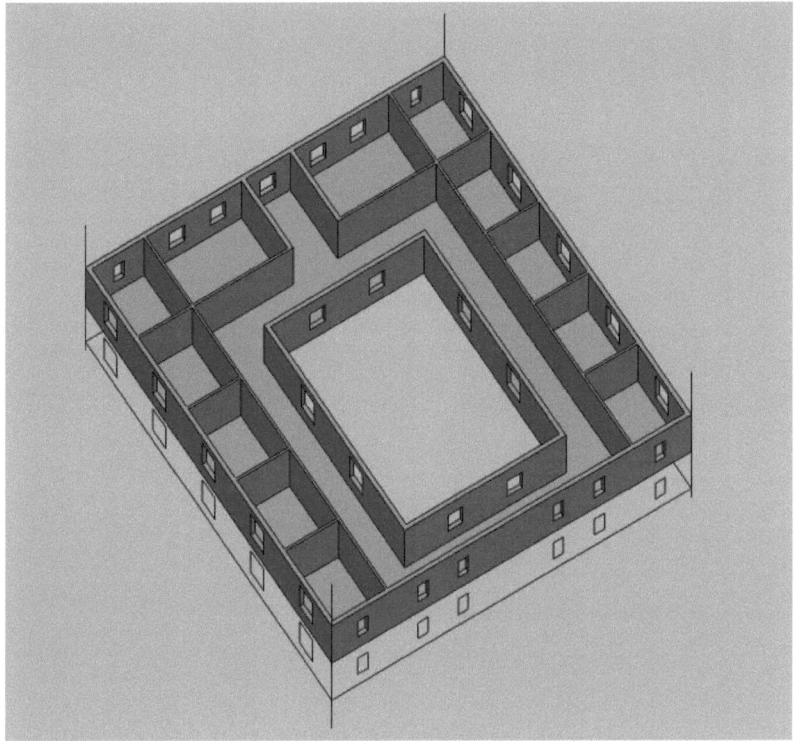

Figure 34: Standard floor of an office building with an unlimited number of identical floors above and below.

4.8.1 Free-float model, windows partly openable

The reduction algorithms outlined in 4.6 and 4.7 were applied to the model displayed in Figure 34. According to 4.7 a zone is selected whose temperature profile is of particular interest. Neighboring zones are added to a list of zones to be left unmerged. In the example displayed in Figure 35 it turns out that the next level of neighboring zones is sufficient to be left unmerged. Thus office rooms on the right are merged as well as the remaining four office rooms on the left hand side.

Windows in the office rooms are kept closed in the simulation whereas windows in the corridor are opened if the inside air temperature exceeds 27 K until the temperature drops below 23 K. The model is configured as a free-float model subjected to meteorological conditions only.

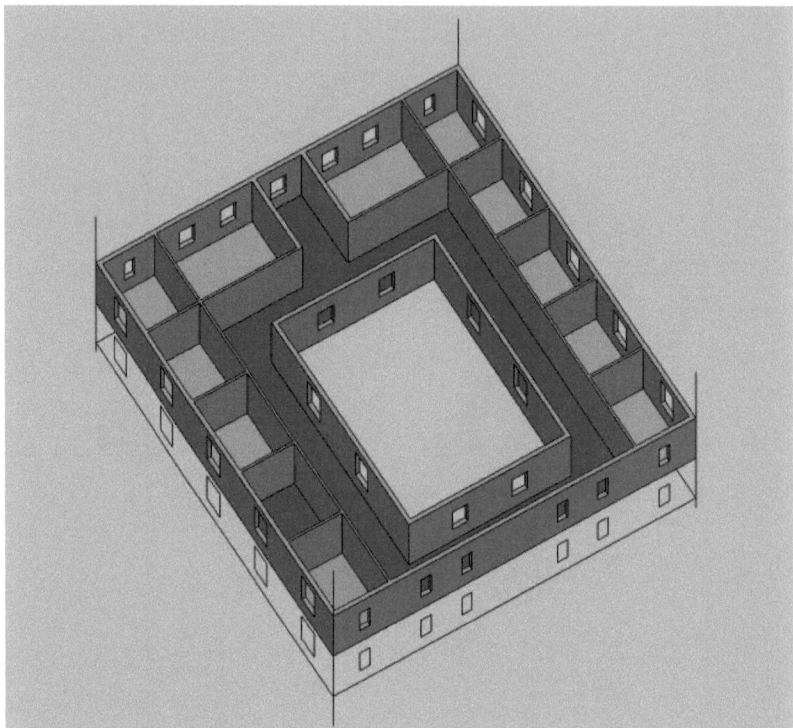

Figure 35: The temperature profile of the zone marked green is of interest. Applying the reduction algorithm according to 4.7 a model is created where only the colored zones are left unmerged.

An annual simulation is run and the resulting temperature profiles of the zone in the corner are displayed in Figure 36. Since the gap between the two profiles is very small they can hardly be kept apart. The average deviation is 0.04 K and the maximum deviation 0.4 K.

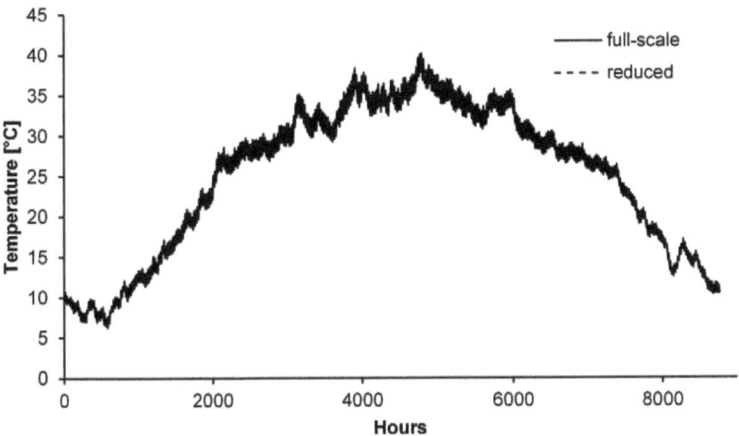

Figure 36: Annual temperature profiles of the zone marked green in **Figure 35** in the context of the full-scale model and of the reduced model.

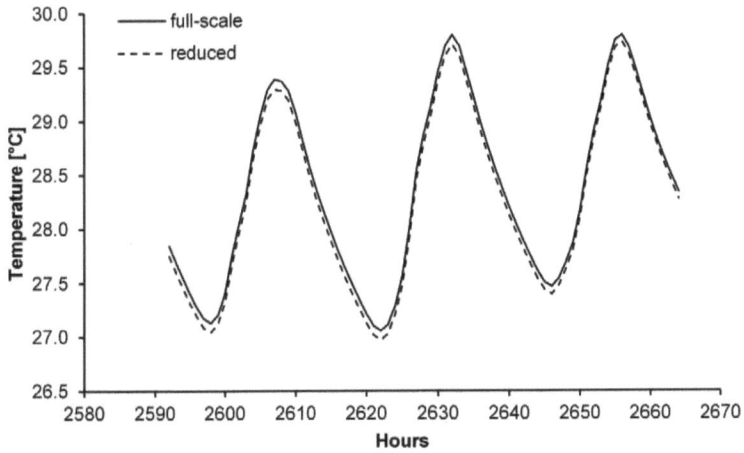

Figure 37: Short term temperature profiles of the zone marked green in **Figure 35** in the context of the full-scale model and of the reduced model.

Figure 37 displays the temperature profile during three days where a relatively large difference in temperatures (compared with the average temperature deviation) between the full-scale and the reduced model can be observed. Even so the results of the reduced model are close to those of the full-scale model.

Model	Order	Duration of simulation (s)	Performance gain (%)	Avg. temperature deviation (K)
Full-scale	133	2028.4	-	-
Reduced	59	1260.4	37.8	0.02

Table 10: Summary of results of a "free-float" simulation of a full-scale model compared with a reduced on according to **Figure 35**.

4.8.2 Ideal heating and cooling systems, windows closed

Zone grouping according to chapter 4.6 returns a scenario where the corridor remains ungrouped and the offices on both sides are merged, resulting into a three-zone model. (Figure 38)

The goal was the calculation of the demand of heating and cooling energy and the relevant loads and the windows were configured to remain closed at all times.

The heating system was configured to emit the ideal amount of heating energy in order to keep the inside air temperature at 21 K as long as meteorological conditions would lead to a further decrease in temperature.

Accordingly the cooling system was configured to withdraw the ideal amount of energy in order to keep the inside air temperature at 25 K as long as meteorological conditions would lead to a further raise in temperature.

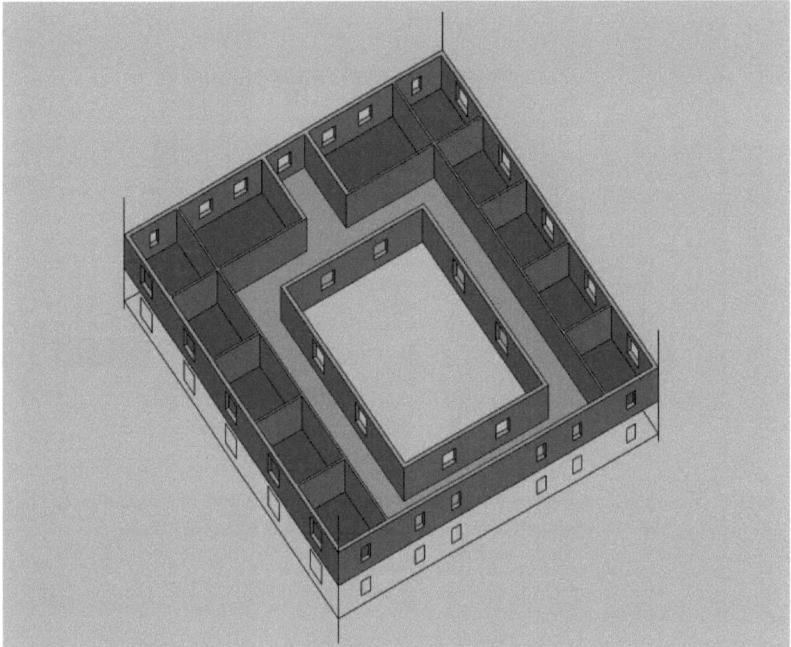

Figure 38: Standard floor of an office building where the zones representing office rooms (blue) are merged.

An annual simulation was run with the model and the temperature profiles of the corridor in the context of the full-scale model as well as in the context of the reduced model according to Figure 38 are displayed in Figure 39. Comparisons of other temperature profiles between the full-scale model and the reduced one seem little meaningful because several zones of the full-scale model are mapped with a single zone of the reduced one.

Even though the goal of the current reduction was the calculation of cooling and heating energy also the temperature profile of the corridor within the reduced model matches the profile within the full-scale model well. (Figure 39)

Again the temperature profile during three days after about one third of the annual simulation is displayed separately. (Figure 40) The relevant cooling loads are displayed in Figure 41.

From the graphical presentation it can be seen that temperatures as well as cooling loads match very well also after about one third of the simulation.

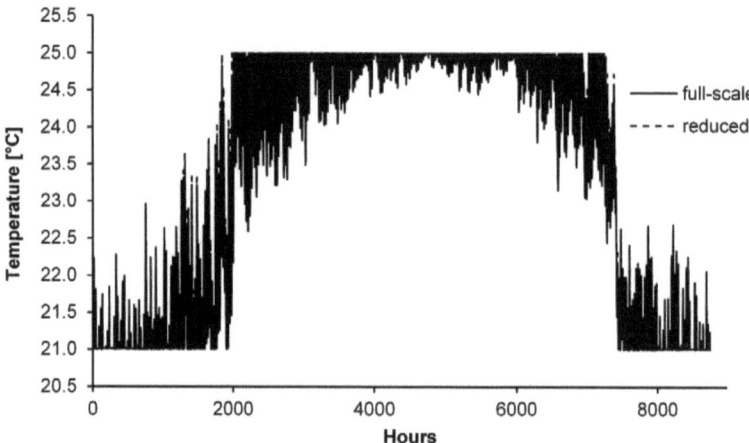

Figure 39: Annual temperature profiles of the corridor within the full-scale model and the model subjected to zone grouping according to **Figure 38**.

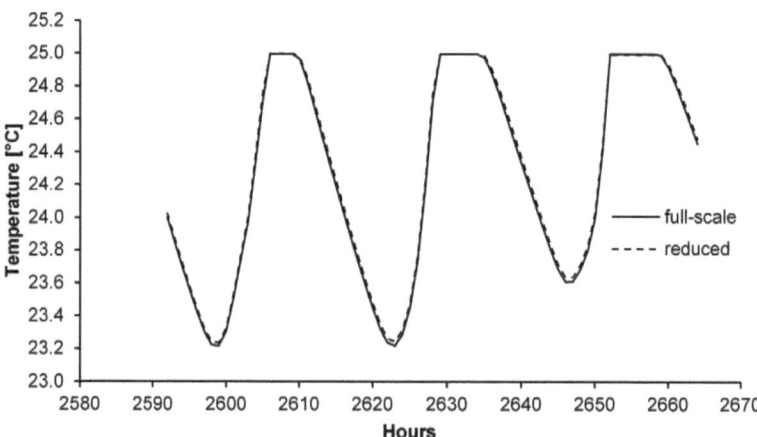

Figure 40: Short term temperature profiles of the corridor within the full-scale model and the model subjected to zone grouping according to **Figure 38**.

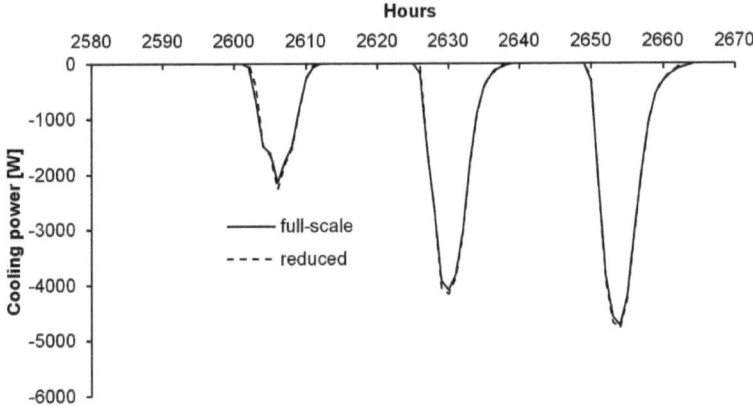

Figure 41: Short term profiles of the overall cooling power matching the temperature profile in **Figure 40**.

In another step the model was reduced even further to a single zone model. Windows were kept closed again and the cooling and heating loads were calculated for an annual simulation.

The results of the simulation are given in the following table:

Model	Order	Duration of simulation (s)	Performance gain (%)	Avg. deviation of overall heating load (%)	Avg. deviation of overall cooling load (%)
Full-scale	133	2619.6	-	-	-
Reduced	37	1214.0	53.6	1.39	0.31

Table 11: Summary of overall results of a full-scale model compared with a reduced on according to **Figure 36**.

4.8.3 Ideal heating, openable windows

Another test is run where the heating system is configured to maintain a minimum air temperature of 21 K. Windows are configured to be opened during the day if the inside temperature rises above 27 K and is above the air temperature. Once they are open they are closed only when an inside temperature of 23 K is reached.

Model reduction is accomplished according to Figure 38, thus a three zone model is subjected to annual simulation.

In another step the model is reduced even further to a single zone model and an annual simulation is run with the same configuration as above. (Ideal heating, windows openable during the day.)

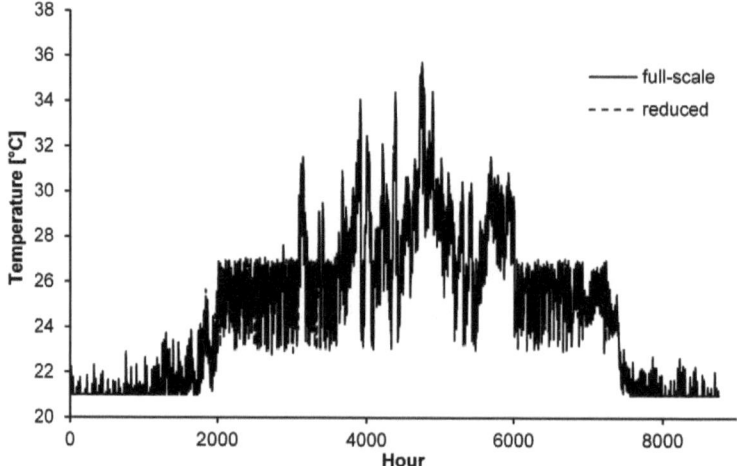

Figure 42: Annual temperature profile of the corridor within the full-scale model and the one subjected to zone grouping according to **Figure 38**.

The annual temperature profiles of the corridor match well for both models, the full-scale one and the reduced one. (Figure 42)

A detailed three days' profile is displayed in Figure 43. Knees in the temperature profile come from windows being shifted from closed to open and back.

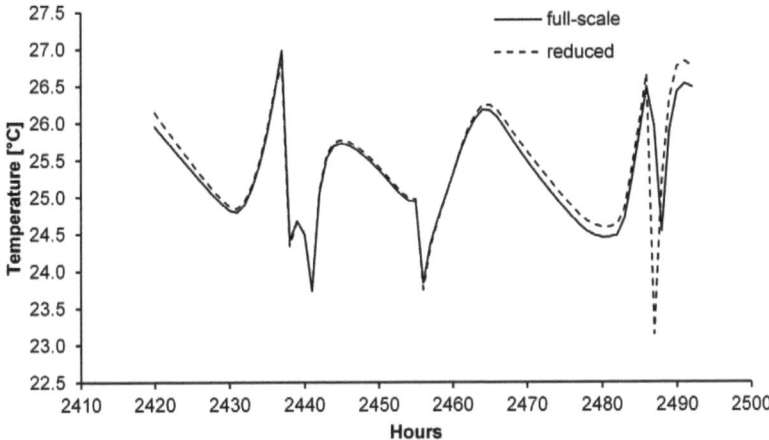

Figure 43: Three days' temperature profiles of the corridor in the context of the full-scale model and of the reduced model according to **Figure 38**.

Model	Order	Duration of simulation (s)	Performance gain (%)	Avg. temperature deviation (K)
Full-scale	133	7422.6	-	-
Reduced	37	1441.6	80.6	0.11

Table 12: Summary of results for the corridor in the context of a model reduced according to **Figure 35**. Windows are openable in summer and heating system is set into ideal operating mode in winter.

4.9 Conclusion

Wall merging is an effective means to reduce the order of a RC model without a loss of accuracy. The algorithm is easy to implement and can be fully automated. Models that result from a transformation of a CAD model into a RC model are likely to dispose of a reduction potential due to wall merging because optimization of RC models is usually not performed immediately. Identical walls for instance are likely to be created separately in a RC model in order to reflect the relevant CAD model the best possible way.

Wall merging can even lead to further model reduction if walls with similar properties are merged instead of just equal ones.

Basing further reduction methods on a periodic steady state reduces the influences e.g. of initial temperatures. This is important for the comparison of the thermal behavior of two zones. However, the calculation of the periodic steady state is not always trivial. A multi-zone model can react sensitively even to small changes in the initial values or excitations. Thus it can be difficult to meet the condition formulated in eq. 4.3.

The time that the conversion to a periodic steady state takes and the number of periods that have to be simulated respectively is obviously closely linked to the initial values that are applied. Permissive zone merging according to 4.5 can lead to a reduced model that allows the quick calculation of the periodic steady state. The resulting temperatures from the periodic steady state of the simplified model can be mapped with the initial temperatures of the full-scale model in order to set the model into a status close to the periodic one. Experiments where zone merging was accomplished even to an extent that multi-zone models where transformed into single-zone models lead to good results regarding the approximation of an initial value temperature.

Two ways of model reduction have been presented. One aims at a fast calculation of the demand of heating and cooling energy and of the required heating and cooling load, the other one at a fast calculation of a temperature profile of one or more particular zones.

Model reduction for energy calculation is based on zone grouping where zones are merged that have a similar thermal behavior. The algorithm can be fully automated and proved to work well in practice. Depending on the size of the gap that is allowed between the temperature profiles of two zones, groups can be kept rather small or include many zones. Particular zones can easily be marked to be left unmerged if they are of special interest or if internal loads or temperature-driven ventilation is expected to spoil a correct result.

Model reduction for the fast calculation of a temperature profile of a particular zone is more complicated than energy calculation-based reduction because the periodic steady state does not only have to be calculated for the full-scale model and the most reduced model according to 4.7 but also for any model set up during the reduction process. Due to the effort that the periodic steady state calculation brings along the reduction process itself may cost some time. However compared with a long term simulation that usually follows the effort of the reduction process itself should not carry too much weight. As with the preceding reduction algorithm also in this case particular zones can be marked in order to remain unmerged.

Both reduction methods can easily be applied with existing RC models. They can be fully automated or completely skipped without a modification of the underlying implementation. Also a switch from automatic to manual selection for zone-merging can easily be accomplished. Overruling automatic reduction methods can be useful if some zones are expected to have a bigger impact on the overall result than others, e.g. due to distinct internal loads or different zone or wall properties.

5 Architecture of the implementation

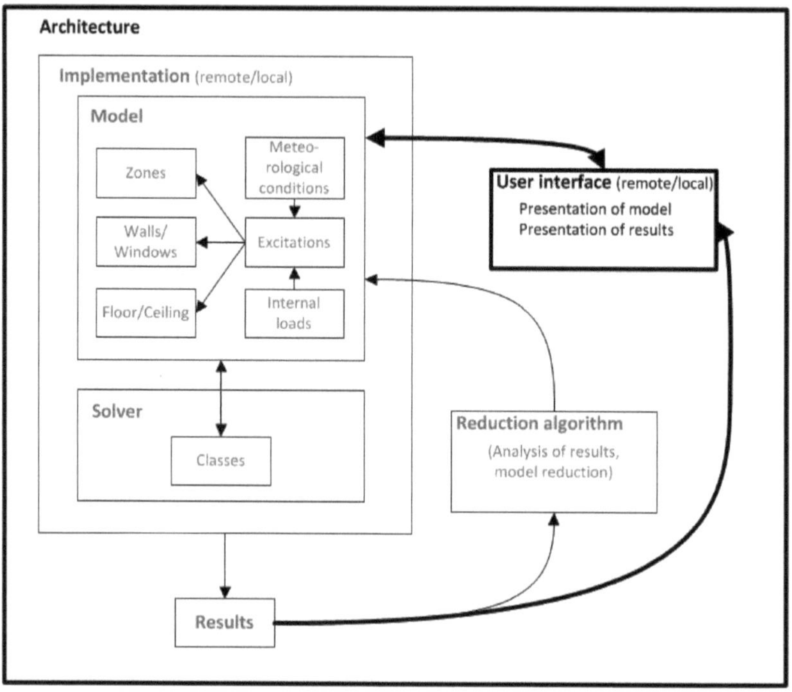

Figure 44: The architecture of simulation system is influenced by the decision about whether a client server or a standalone implementation is accomplished.

Figure 44 shows that the architecture is wrapped around the implementation of the model, control systems and the solver. Intended update cycles and ways of distribution to the public can be relevant for the decision about the architecture, for instance if the simulation engine should remain at the server-side and only the client part is distributed.

The decision of the architecture of a software tool is often driven by the availability of computing power and the need for communication between the individual users. Sometimes also the availability and dependability of information at different nodes of a network may play an important role. This is often the case with web applications or simple websites used to distribute information over the internet. Byrne et al. (2010) give an overview of the different types of architecture of web-based building simulation software. Yahiaoui et al. (2008) use SOAP for runtime coupling of simulation and control systems. Isikdag (2012) presents concepts for data exchange based on AJAX for visualizing purposes and on REST for modeling purposes. Also Meyer zu Eissen and Stein (2006) have investigated web-based simulation services. They suggest a SOAP-based add-on for an existing simulator.

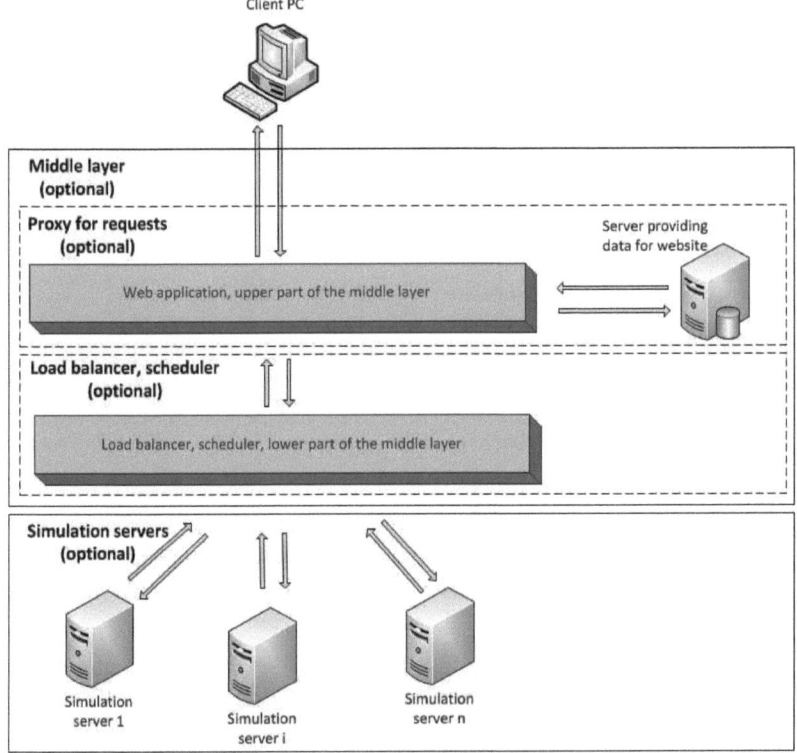

Figure 45: Overview of an architecture with three optional layers. Missing layers have to be replaced by the client PC

Figure 45 contains a schematic presentation of the individual layers that the application setup is composed of. In a distributed environment the lowest layer holds a server cluster designated to host the simulation engine and to run the simulations. The number of servers is chosen with respect to the number of users in order to avoid bottlenecks in production mode. It depends on the average duration of simulation and on the expected number of simulations requested per time unit. A reasonable approach could be to define the probability that an arbitrary request can be served immediately. The expected standard deviation of the duration of a simulation run from the average duration and the expected number of requested simulations can be used to calculate the number of servers to achieve the defined level of service.

The middle layer consists of two sub layers. The upper one hosts a web application serving a website with information, checking login credentials and passing requests on to the next layer.

The lower sub layer is designed as a load balancer and scheduler. It keeps track of the users' requests and the servers' workloads. If all simulation servers are busy it can set up an order for serving pending requests and pass them on to the simulation servers.

The middle layer including the load balancer and the scheduler as well as the proxy accepting the users' requests can be omitted. In this case the users would send their requests directly to the simulation servers.

In a local standalone scenario the user and his PC would have to undertake the tasks of the middle and the server layer. In fact a typical user would not think of a local installation as a distributed layered application. Scheduling and load balancing would happen intuitively as the number of parallel simulations is usually limited to the number of CPU cores.

Figure 46 holds a sequence diagram of a setup without middle layers. In this case there is a single request-response cycle between the client and the server. The disadvantage is that in a single-threaded client environment the control flow is blocked while the simulation is executed.

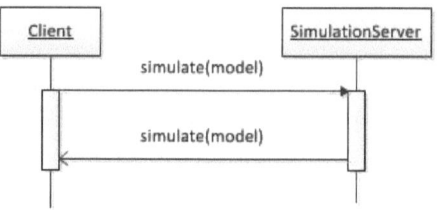

Figure 46: Overview of a synchronous request, response cycle. There are no middle layers in this setup

Figure 47 displays a scenario where a proxy takes all client requests, returns expected simulation durations and asynchronously passes the request on to the simulation engine.

Any client requests between the start of the simulation and its finish are served by the proxy. In this way also multiple requests for the same simulation can be intercepted and handled appropriately.

When the simulation results are ready the proxy acts as a client on its part. It passes them on to the client by calling a callback function or a web service (`deliver(Results)`) located on the client side.

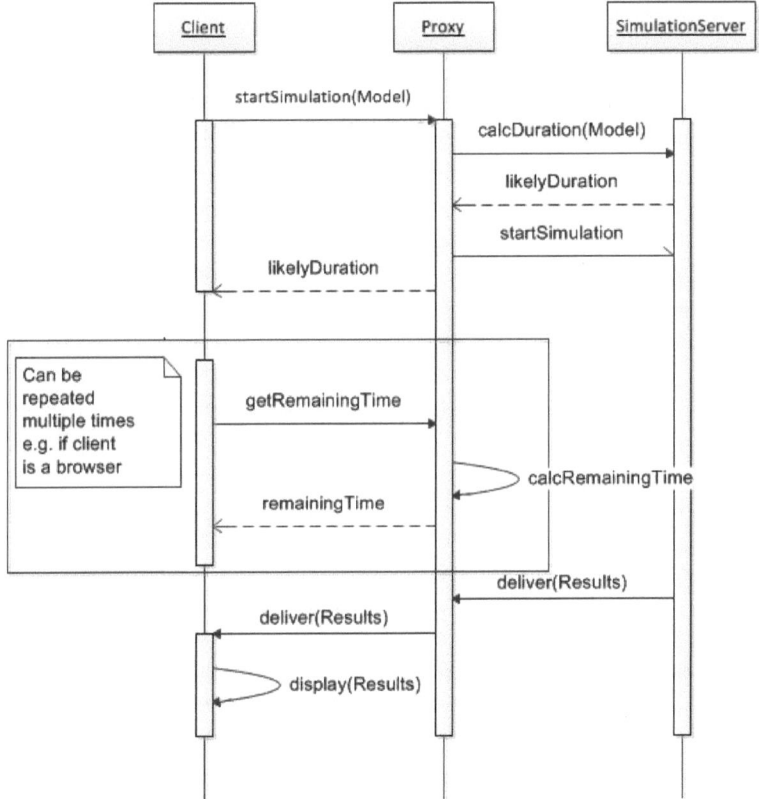

Figure 47: Overview of a setup including a layer between the client and the simulation servers. The client disposes of a callable function or web service itself.

Figure 48 displays a similar setup as Figure 47 until the point where the simulation results are ready. The difference between the two setups is that in setup of Figure 48 the client cannot act as a server and provide a callback function. This complicates the implementation because it has to poll continuously for the simulation results.

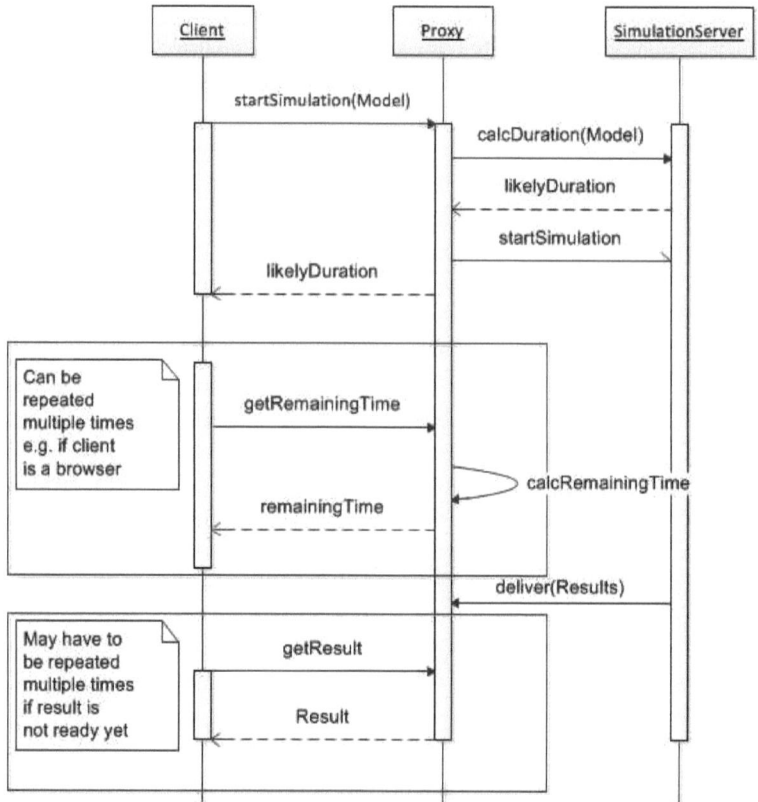

Figure 48: Overview of a setup including a layer between the client and the simulation servers. The client does not dispose of a callable function or web service itself. Instead it has to poll for the result.

In the following a short overview of the pros and cons of a service-orientated and a local stand-alone architecture will be given.

5.1   Computing power

Computing power is an important factor for running a simulation because it is proportional to the speed of the execution.

If a simulation is distributed, installed and run on an end user's machine no guarantees can be given regarding the type and power of the hardware that is used. On the other hand every user is in control of scheduling his or her simulations and no bottleneck can occur to one user because of costly simulation tasks of another user. This is not the case with a service-orientated architecture. If there is a set of servers executing the simulation tasks of all end users not enough CPU power may be available for the individual simulation run. A particular user may have to wait for his turn if all servers are busy. Byrne et al. (2010) mention the loss of speed as one the disadvantages of web-based simulation.

The hardware used by an end user may not be well suited to running a building simulation because users usually choose their machines with other targets in mind or because the relevant machines are simply too old. In this case the performance may suffer. In case of a service-orientated architecture, however, machines can be purchased and setup the best possible way with respect to the performance of simulations.

So with a limited number of users a service-orientated architecture can be advantageous because the hardware can be chosen and setup with the best possible efficiency regarding the simulation engine. However, if the number of users increases and the computing power available on the server side is not sufficient bottlenecks can occur.

5.2 Distribution and installation process

The installation process of a software tool can be very simple or rather tedious depending on the components that the complete tool is made up of. If the operating system does not come with the interpreters and runtime engines that are needed they have to be included with the installer. Many programming frameworks provide guidelines or supporting tools for creating installers for different platforms. Nevertheless slim installers are favorable because they can more easily be distributed and consume less resources than big ones.

Standalone tools have to be distributed including the engine that the simulation is to be run with whereas in service-orientated architectures only the client part has to be shipped. In some cases, e.g. if a browser is used as user interface, no client installer has to be delivered at all. Meyer zu Eissen and Stein (2006) present an approach where they provide an applet for download and usage with common browsers. Nowadays an applet may not even be necessary anymore because browsers support more sophisticated JavaScript-based user interface design these days than a couple of years ago. (The usage of an applet requires a Java plugin for a browser which may not always be available at an end user's machine.) In both cases, JavaScript-based user interface design and an embedded applet, the integration with an existing simulation engine can be achieved via a web application acting as a middle layer between the user and the server hosting the simulation engine.

Another major advantage of service-orientated architecture concerns the distribution of updates because changes of the server part usually have no effects on the parts distributed to the users.

5.3 Network traffic

Building simulations as such do not cause network traffic, neither in a local standalone architecture nor in a network service-orientated one. In a service-orientated architecture, however, requests are sent from the users to the server and responses from the server to users. In case of the implementation outlined in the thesis at hand typical requests and responses are about 10 kB per zone. In an average network infrastructure this order of traffic can be handled without any problems.

5.4 User interface

RC models have a huge advantage compared with other types of models: There is a clear mapping between the construction elements and the zones on the one side and the individual equations on the other side. All parameters have a clear physical meaning.

The reduction of the equation system can therefore be accomplished more easily than e.g. with black-box models because the impact of particular parameters can be estimated from the beginning.

Another significant advantage concerns the design of the user interface: The mapping from CAD or 3D models with RC models is obvious at all times. User interfaces for CAD and for RC models can therefore be similarly designed. User interfaces can come with an intuitive handling also for users who are not yet familiar with building simulations.

5.4.1 Requirements for user-friendliness

There is no general-purpose definition of what a building simulation system has to consist of. The implementation is not necessarily restricted to the simulation engine as such where the underlying equation systems are solved. On the other hand there are no regulations that demand the delivery of a GUI together with the implementation of the simulation engine. The number of available building simulation tools may well be as high as the number of different combinations of implementations of solvers and GUIs.

The decision of what is provided with a tool depends on many aspects. One of them is the target group of users that a tool is intended for. The most obvious users are the developers themselves. The idea for the development of a simulation tool often originates from a problem that the developer himself is confronted with and for which a solution is needed. The distribution of a tool is often not intended at the beginning of a development phase. Only when it turns out that an implementation serves the original purpose as well as expected or even better the distribution is initiated. At that point, however, the complexity of an implementation has often grown to an extent where people who have not been involved in the actual development process have difficulties in even just learning to use it. Simplifications in the sense of user friendliness have become difficult at that point, first because the intention to launch the distribution has come late and decisions from an early stage of the project which foil user friendliness cannot be made undone and second because developers have become used to their own product which makes them unaware of the difficulties that may occur to others who are not as involved as themselves. Solutions for these troubles may be hard to find or consist of unsatisfying workarounds at this point.

5.4.1.1 Architecture driven considerations

A user interface has to accomplish several tasks. It shall not only provide access to the simulation engine in order to pass models for simulation purposes but it also should be capable of displaying a presentation of the models as well as of the results returned after a successful simulation run. A typical user is not interested in whether the simulation engine is setup on a remote computer or locally. He rather wants a user-friendly presentation of the model and the results. In fact in times of SOAP, REST and other high level network protocols the user interface is rather independent of the location of the simulation engine. So the architecture as such has no direct impact on the user interface. Where the architecture of the simulation tool does play a role is if access without an installation process shall be granted, that is, e.g. the browser should act as a user interface out of the box. This is a requirement that can only be met if no installation procedures (additionally to setting up a browser) need to be completed at a local computer.

5.4.2 Browser

Using the browser as a user interface is a tempting approach because a browser is available at almost any machine. In a distributed environment it introduces the possibility to request simulation runs from practically anywhere anytime. The downside of the approach is that the original purpose of the browser is not to set up highly sophisticated building models and send requests to simulate them to the relevant servers but to display rather static information from the internet. Even with modern technologies like JavaScript and HTML5 some tweaks may be necessary to create a user-friendly browser-based interface. Another disadvantage is that the web application (see also Figure 45) acting as a proxy cannot be omitted. The effort of developing and setting it up may grow rapidly especially in multi-user environment.

5.4.3 Handmade GUI

Another type of user interface would be based on an OS-dependent GUI implementation. E.g. MS Visual Studio provides libraries and controls to create elaborate GUIs, including tabbed controls, controls for tables and lists, file management and many more. The disadvantage of these types of GUI is that the development process is costly and time-consuming and that the required development tools are usually rather expensive.

See et al. (2011) for instance have started to develop a GUI for EnergyPlus. They do not intend to replace existing applications, capable of building information modeling (BIM) but they rather offer an addition to it. Simple model generation based on template shapes is possible as well as the import of formats like IFC and gbXML. They also implement rule based checks in order to help the user avoid the input of incorrect or inconsistent model configuration data.

5.4.4 MS Excel

Since the interface to the simulation engine within the implementation at hand consists of a SOAP web service any client capable of sending a SOAP request can be created. Due to its features as a spreadsheet program and its wide dissemination MS Excel is chosen for the prototype implementation of a client user interface. One of the features of MS Visual Studio is to create plugins for MS Office components. A plugin including a web service client was created in order to send simulation requests to the simulation engine. So users can organize their models in spreadsheets or load them into spreadsheets from custom databases and send simulation requests at the click of a mouse out of Excel.

Figure 49 displays a screenshot of a prototype of such an MS Excel file. On the left hand side there is a service reference to the relevant SOAP service. The service can be accessed by simply creating an instance in the underlying Code. The request is sent by clicking the "*Simulate*" button. Subsequently the data representing the relevant model is collected from various spreadsheets. It is copied into the objects forming the request. Then the simulation is triggered by sending the request to the server. When the simulation is finished the results are returned to the Excel client and the data from the relevant response objects are copied into the result spreadsheets.

A problem related to the client nature of the Excel plugin comes from the synchronous specification of the SOAP protocol. When a request for simulation is sent to the server the response is only returned when the simulation is finished. In the meantime any user interaction is blocked because the web service client is waiting for the response of the server and Excel does not allow any user input during that phase. This is not a convenient solution in particular because the user needs to be informed about the remaining time until the simulation results are ready.

Architecture of the implementation

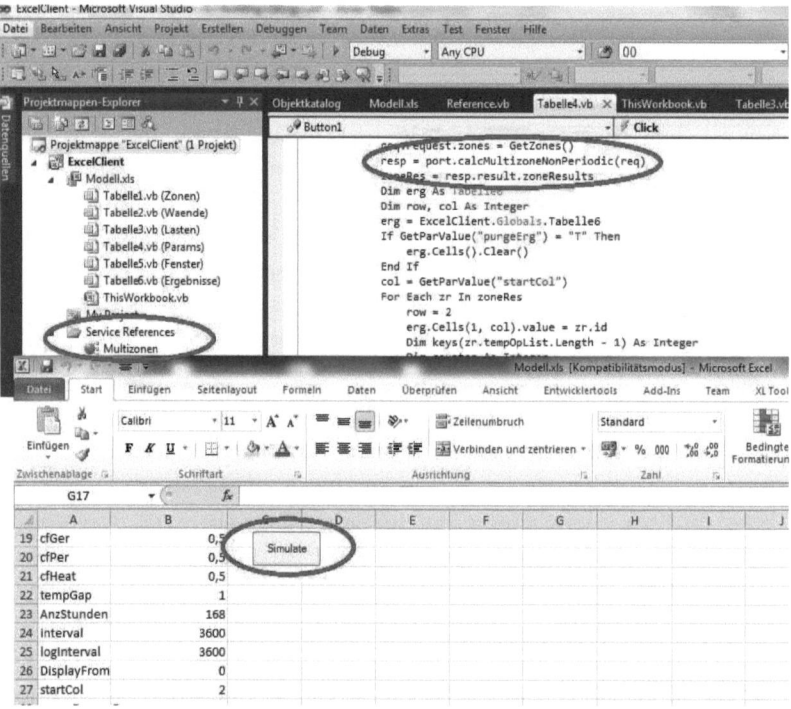

Figure 49: With the help of a plugin MS Excel can be used as a web service client. The code for accessing the web service is written in MS Visual Studio. A representation of the model is held in several spreadsheets of an Excel file.

6 Conclusion

6.1 Model

A RC model has been chosen for simulation purposes. RC models have advantages compared with neural network or finite-volume-based ones regarding many aspects. Some of them have been mentioned in 1.1. The most decisive ones include performance advantages e.g. compared with neural network or finite-volume models. Also a clear mapping between model specifications and configurations in the user interface and the underlying equation system is most important for creating a user-friendly application.

A *3R2C* sub-model has been used for the construction elements. In literature this type of model is often considered to provide a good tradeoff between result accuracy and simulation performance. Zone temperatures are assigned an internal heat capacity. Walls, ceilings and floors that are attached to a particular zone or serve as partition walls between two zones (and thus are attached to both zones) represent the central elements of the RC model. Every construction element produces two equations and every zone one within the underlying equation system. The number of equations can be reduced on condition of symmetry properties of internal walls.

Windows are an important factor for setting up a meaningful model. While the thermal mass of a window is negligible compared with much heavier construction elements the impact of surface temperatures on the overall profile can be significant. Due to the negligible thermal mass additional equations can be saved. Instead the surface temperatures are calculated by taking into account both, the radiative temperature and the air temperature on both sides of the window. The conversions for an explicit representation of T_{Rad}, the interior radiative temperature, which is needed for an efficient implementation, are outlined in 2.3.

Solar gains are strongly linked with the calculation of shadow mapping. Shadow mapping can basically be implemented in a traditional or in a GPU-based way. The traditional way requires projecting the coordinates of the relevant polygons unto a projection plane. A depth sorting according to Weiler's and Artherton's (1978) algorithm has been accomplished in order to set up an order in which the polygons are clipped with one-another.

A goal of the outlined approach was to allow the operation of the relevant component on a separate machine in order to distribute CPU loads. A proposal for the architecture of the implementation in web service was given not least in order to create a simple interface.

GPU-based shadow mapping represents a more recent approach to calculate shadows. It is based on the exploitation of GPU power and aims at real-time shadow calculation. Problems can occur due to the complicated access of GPUs as state machines as well as due to the accuracy of results. GPU-based shadow mapping may require multiple displacements of the viewport which may cause unexpected problems related to the geometry of the current scenario.

Meteorological conditions that have to be taken into account in a building simulation include among others outside temperature and direct and diffuse irradiation. Whereas solar gains through windows are calculated separately with respect to the shadowed parts of a building Hagentoft (2001) presents a way to calculate an equivalent temperature that represents the impact of irradiance and air temperature in a single value. It is applied with the calculation of exterior surface temperatures of the model.

6.2 Implementation

The choice of a programming platform represents the most far-reaching decision regarding the implementation of a simulation tool. Several requirements have to be fulfilled by a programming platform. The most important ones have been listed in 3.1 and include support for object-orientated programming and bindings for efficient solvers to name two of them. C++, Matlab and Python have been compared regarding these requirements. The three of them can be suitable for the implementation of a simulation tool. The decision was made to use Python because of its versatility, among other advantages.

The main effort consists in transferring the building model into the equation system that is passed to the solver. The application has to allow for the particulars of the construction elements and the zone properties as well as for the creation of the objects associated with internal loads, meteorological conditions, etc.

Some of the CPU load can be removed from the simulation engine by transferring certain calculation procedures to other machines. The most important ones include solar gains with respect to shadowing and the equivalent temperature representing the meteorological conditions. Running these calculations on different machines prior to the actual simulation can bring about a significant performance gain.

Usually the creation of a CAD model is accomplished before a RC model is set up. The CAD model often serves as a rough template for the RC model. However the type of information that is represented by CAD and RC models respectively differs significantly. Thus an automated translation from CAD into RC models is difficult.

There have been efforts related to automatic model setup but so far none of the available solutions has reached a mature state.

Heat capacity values are subject to comprehensive research. Approaches range from mapping measurement data with model properties to analytic solutions. The approach that was used with the model at hand is based on a Fourier transformation of a sinusoidal excitation.

The ODE solver represents the core of a simulation engine. It should be chosen carefully with respect to the characteristics of the equation system. Euler's method is often used in building simulation. The step-size of the solver is chosen explicitly with Euler's method. It has been shown that it is vital for the tradeoff between accuracy and performance. If the step-size is set to a fixed value the accuracy results automatically from the characteristics of the equation system. Variable-step solvers can have advantages compared with fixed-step ones regarding performance and error tolerance. They allow the configuration of an error tolerance. In this case the step-size is calculated to comply with the given tolerance.

LSODA represents an implementation that automatically selects one out of two ODE solvers with respect to the characteristics of the equation system. It is recommended if no stringent conclusions can be drawn regarding the stiffness of the equation system. Thus it was mostly used with the implementation at hand.

RC models are often used with the development of MPC or control systems in general. Modern control strategies can even allow for weather forecasts in order to optimize energy consumption in buildings. The model at hand represents a good basis for trying out different control strategies.

The control strategies applied to the current simulation have been selected with the objective to reflect a realistic behavior of the model. Depending on the type of excitation a time-driven or a temperature-driven strategy has to be implemented.

Ventilation control for instance can usually not be set up to act uniquely time-driven or uniquely temperature-driven. A mixed strategy has been applied to allow for the presence of people just as well as for a sudden fall in outside temperature.

In contrast control strategies for shading and internal gains have not been set up in a mixed mode.

Heating and cooling are of particular interest in building simulation. The demand of heating energy contributes a major part to the overall energy performance and the required heating capacity is of particular interest for the dimensioning of the heating system. An explicit representation of the required heating energy has been developed and a strategy has been evolved that allows keeping the zone temperature within an upper and a lower bound. The conversions that have been applied lead to performance gains and to exact results.

6.3 Reduction algorithm

Many researchers agree that also with RC models reduction methods make sense in order to run long term simulations of multi-zone building models in a shorter amount of time.

The most important fields of research regarding performance tuning of RC models include the variation of the time step and model reduction by fewer nodes. Model reduction is often achieved by associating bigger parts of the building with fewer nodes and adapting the calculation of the lumped capacity. Other reduction methods are based on mathematical approaches like single value decomposition.

Reduction methods can be based on the reduction of a full-scale model or on the creation of simplified model from scratch. The thesis at hand presents a reduction algorithm that is based on the relevant full-scale model. The approach allows identifying the parts of the model that spoil the performance but are not needed for an adequate accuracy of result.

Model reduction is reached in several steps:

- Exploitation of symmetry properties of partition walls.
- Exploitation of symmetry properties of internal walls in periodic zones.
- Aggregation of similar walls.
- Merging of zones with similar temperature profiles under steady state conditions.
- Iterative identification of zones to be left unmerged based on the distance to the zone(s) of interest.

Model reduction by aggregation of similar walls is based on the interpretation of individual walls as fragments of the original wall that they have been derived from. If heat resistance, heat capacity and adjacent zones match walls are merged and the order of the equation is reduced accordingly.

The prerequisite for the identification of zones with a similar impact on the thermal behavior of a building is the calculation of a periodic state. Thus the influences of the initial conditions can be reduced. The calculation of the periodic state is based on the reduction of the model to a single zone itself.

Once the periodic state is reached two ways of zone merging can be applied. In one case the algorithm aims at the calculation of the demand of heating and/or cooling energy. Mergeable zones are identified by the similarity of their temperature profiles. If the difference between the temperature profiles is small the relevant zones are considered to be mergeable because the relevance of distinct zones for energy calculation is likely to be negligible.

In the other case the goal is to calculate the temperature profile of a particular zone and an iterative approach is applied. The periodic state is calculated for the full-scale model and for a maximally reduced one (i.e. with all zones merged). Then step by step the zones adjacent to the zone of interest are added to the list of unmerged zones and the periodic state is calculated again. Then the temperature profiles of the zone of interest are compared. The model reduction is ready when the difference between the temperature profile within the full-scale model and the one within the reduced model are small enough. The algorithm aims at a precise calculation of the temperature profile of a particular zone within a reduced model.

The algorithms proved to work well in practice. They were applied to several models and achieved significant performance gains at an acceptable loss of accuracy.

One of the major advantages of the suggested reduction algorithm is that some steps can be applied without any loss of accuracy. Wall merging and the reduction of the relevant ODE system due to symmetry properties of internal walls can often be applied if the relevant RC model is directly derived from a building model.

Another advantage is related to the automation potential of the algorithm. All steps can be fully automated. However manual intervention is still possible if for instance some zones shall be excluded from undergoing certain reduction steps.

Manual intervention can also be useful if temperature profiles of more than one zone are needed. It is possible to tag several zones to remain unmerged without having to run the simulation several times.

6.4   Architecture of the implementation

A proposal for the architecture of a simulation system is made. In a distributed application a client and a server is needed. A middle layer is optional as long as the number of users is limited. End users represent the clients of the application. The middle layer consists of a web application that can hold guidelines and documentations or undertakes the user administration. It can be omitted if none of these tasks has to be fulfilled. The load balancer is the part of the middle layer that accepts requests from clients and distributes them to one of the available servers. When the request has been finished and the response is ready the load balancer passes it on to the client. It can also serve to calculate response times and to server poll requests from clients.

If the middle layer is omitted the clients send the requests directly to the servers. This is manageable if the number of users is small. Otherwise the distribution of request has to be taken care of separately by the load balancer.

Computing power is the one limited resource that causes the need for performance tuning. While simulations can be run in parallel if they are distributed to multiple machines or multiple cores respectively it is currently not possible to speed up a particular simulation by multiplying CPU cores. This is an active field of research but so far no breakthrough has been achieved in exploiting parallel computing with coupled ODE systems.

Even with respect to the number of simulations run in parallel a simulation system can run out of computing power. Depending on whether a PC hosts the simulation system itself or acts as a client that just sends a simulation request to the server the availability of computing power may vary. In a client-server environment hardware facilities can be better adjusted with the needs of a simulation engine but if many users send requests simultaneously a bottleneck can occur.

Client-server solutions have a clear advantage over standalone packages regarding the distribution and installation process. The simulation engine remains at the server thus only the client part has to be distributed which shrinks the installer significantly. If the browser is used as a user interface no installation is needed at all.

Updates in most cases only need to be applied to the server. Clients are usually not affected unless changes of the user interface are distributed.

The amount of available network bandwidth has constantly grown during the last years. Network traffic related to building simulation rarely exceeded some kB during tests. Thus network bandwidth is unlikely to pose a problem with client-server architecture.

The user interface is a particularly important component if a distribution to the public is intended because it is vital for the acceptance by the users. Users tend to stick with tools they are familiar with. Gaining enough knowledge to be able to use a new tool always requires overcoming inhibitions. This is why intuitive user interfaces and simple operating modes are often characteristics of success stories on the market.

Developers run the risk of disregarding user interfaces in particular if the idea to publish a tool only comes at the end of a development cycle. The user-friendliness of an interface often lies in the eye of the observer but what seems to be intuitive to a developer need not necessarily be handy for user.

Fortunately in times of SOAP, REST, etc. the architecture is not essential for the user interface anymore. Remote procedure calls are well supported by most programming languages. Using the browser as a user interface avoids complicated setup procedures because a browser usually comes with the operating system. Theoretically even in a local environment it could be used as a front end. However also in times of JavaScript or flash a fancy browser-based GUI may be difficult to design.

This is not the case with a handmade GUI. It can be fully adjusted with the underlying application. However a handmade GUI requires the creation of an installer that must be distributed to the users. Also every update has to be packaged and distributed.

The user interface that is suggested with in the thesis at hand is based on a MS Excel plugin. MS Excel has several advantages regarding the usage as a Web Service client:

- It can be used for auxiliary calculations for instance regarding solar gains or geometric dimensions if needed.
- It is very common among engineers and scientists, thus only the plugin has to be distributed.
- The inhibition level is rather low. Users are more likely to try out new tools if they only come with small changes compared with something they are already familiar with.

The usage of MS Excel as a client also has some disadvantages regarding the synchronous nature of SOAP: If a simulation is started directly from an Excel client (and the middle layer that organizes requests from multiple users is missing) the client is blocked until the simulation is finished. This is not acceptable if a simulation run lasts longer than a couple of seconds. Thus setting up a middle layer will be unavoidable in many cases and a polling mechanism has to be implemented according to Figure 48.

6.5 Future work

Research regarding building simulation in general and RC models in particular is far from being finalized. "Building simulation" as a search string in a popular internet search engine currently returns 75 Mio results and many of them have a scientific background.

Thus it is difficult to pick a single topic that seems worth undergoing further research. However regarding performance tuning an interesting topic of research shall be mentioned anyways. Zhou et al., (2011) have experimented with exploiting GPUs in order to parallelize algorithms of ODE solvers. They made progress in the field of bioinformatics but their problem statement is different than building simulation related ones. Thus also the solution cannot be directly transferred into building simulation.

GPU-related research has become more and more active during the last years. In the future it may lead to groundbreaking results with simulation engines in general and building simulation in particular.

Another point that seems worth further investigations is the calculation of the periodic steady state as a starting point for reduction algorithms. The elimination of the impact of the initial values compares with the computational power effort that is required for the calculation of the periodic steady state. Unfavorable conditions can lead to a substantial prolongation of the calculation of the periodic steady state. Thus research regarding the determination of a state of a model that also qualifies for applying further reduction methods but requires less computational effort would make sense.

7 References

7.1 Literature

Achterbosch, G.G.J., de Jong, P.P.G., Krist-Spit, C.E., van der Meulen, S.F., Verberne, J., 1985. The development of a comvenient thermal dynamic building model. Energy and Buildings 8, 183–196.

Alexander, D., Felsmann, C., Strachan, P., Wijsman, A., 2008. International Energy Agency Building Energy Simulation Test and Diagnostic Method (IEA BESTEST) Multi-Zone Non-Airflow In-Depth Diagnostic Cases: MZ320–MZ360.

Antoulas, A.C., Sorensen, D.C., Gugercin, S., 2001. A survey of model reduction methods for large-scale systems. Contemporary mathematics 280, 193–220.

Arasteh, D., Selkowitz, S., Apte, J., LaFrance, M., 2006. Zero energy windows.

Bueno, B., Norford, L., Pigeon, G., Britter, R., 2012. A resistance-capacitance network model for the analysis of the interactions between the energy performance of buildings and the urban climate. Building and Environment 54, 116–125.

Byrne, J., Heavey, C., Byrne, P.J., 2010. A review of Web-based simulation and supporting tools. Simulation Modelling Practice and Theory 18, 253–276.

Camacho, E.F., Bordons, C., Camacho, E.F., Bordons, C., 2004. Model predictive control. Springer London.

Candanedo, J.A., Dehkordi, V.R., Lopez, P., n.d. A control-oriented simplified building modelling strategy, in: 13th Conference of International Building Performance Simulation Association.

CEN, 2007. EN ISO 13786 Thermal performance of building components – Dynamic thermal characteristics – Calculation methods. CEN, Brussels.

Carslaw, H. S., Jaeger, J. C. 1950. Conduction of heat in solids, Repr. . - Oxford : Univ. Press, 1950. - VI, 386 S.

Clarke, J.A., 2001. Energy simulation in building design. Butterworth-Heinemann, Oxford

Deng, K., Barooah, P., Mehta, P.G., Meyn, S.P., 2010. Building thermal model reduction via aggregation of states, in: American Control Conference (ACC), 2010. pp. 5118–5123.

DIN EN 13363-1:2007-09 Solar protection devices combined with glazing -

Calculation of solar and light transmittance - Part 1: Simplified method; German version EN 13363-1:2003+A1:2007

Dobbs, J.R., Hencey, B.M., 2012. Automatic model reduction in architecture: A window into building thermal structure. Fifth National Conference of IBPSA-USA.

Dos Santos, G.H., Mendes, N., 2004. Analysis of numerical methods and simulation time step effects on the prediction of building thermal performance. Applied Thermal Engineering 24, 1129–1142.

EnergyPlus. <www.eere.energy.gov/buildings/energyplus>, U.S. Department of Energy (DOE).

Foucquier, A., Robert, S., Suard, F., Stéphan, L., Jay, A., 2013. State of the art in building modelling and energy performances prediction: A review. Renewable and Sustainable Energy Reviews 23, 272–288.

Fraisse, G., Viardot, C., Lafabrie, O., Achard, G., 2002. Development of a simplified and accurate building model based on electrical analogy. Energy and buildings 34, 1017–1031.

Freire, R.Z., Mazuroski, W., Abadie, M.O., Mendes, N., 2011. Capacitive effect on the heat transfer through building glazing systems. Applied Energy 88, 4310–4319.

Georgios N.L., Konstantinos F.S., Georgios I.G., Dimitrios V.R. 2013. SRC: A systemic approach to building thermal simulation. Proceedings of the IBPSA Building Simulation 2013 Chambery, 2013

Gladt M., Bednar T., 2013a. A fully automated calculation of shadow casting with matrix-based coordinate transformations and polygon clipping. 13th Conference of IBPSA 2013.

Gladt M., Bednar T., 2013b. A new method for the calculation of the sky view factor for non-rectangular surroundings. 13th Conference of IBPSA 2013.

Gouda, M.M., Danaher, S., Underwood, C.P., 2002. Building thermal model reduction using nonlinear constrained optimization. Building and Environment 37, 1255–1265.

Goyal, S., Barooah, P., 2012. A method for model-reduction of non-linear thermal dynamics of multi-zone buildings. Energy and Buildings 47, 332–340.

Hagentoft, C.-E., 2001. Introduction to building physics. Lund: Studentlitteratur, ISBN 91-44-01896-7..

Hazyuk, I., Ghiaus, C., Penhouet, D., 2012. Optimal temperature control of intermittently heated buildings using Model Predictive Control: Part I – Building modeling. Building and Environment 51, 379–387.

Huang, Y.J., Brodrick, J., 2000. A bottom-up engineering estimate of the aggregate heating and cooling loads of the entire US building stock.

Hudson, G., Underwood, C.P., 1999. A simple building modelling procedure for MATLAB/SIMULINK, in: Proceedings, International Building Performance and Simulation Conference, Kyoto.

Hviid, C.A., Nielsen, T.R., Svendsen, S., 2008. Simple tool to evaluate the impact of daylight on building energy consumption. Solar Energy 82, 787–798.

Isikdag, U., 2012. Design patterns for BIM-based service-oriented architectures. Automation in Construction 25, 59–71.

Jiménez, M.J., Madsen, H., Andersen, K.K., 2008. Identification of the main thermal characteristics of building components using MATLAB. Building and Environment 43, 170–180.

Jones, N.L., Greenberg, D.P., Pratt, K.B., 2012. Fast computer graphics techniques for calculating direct solar radiation on complex building surfaces. Journal of Building Performance Simulation 5, 300–312.

Jones, N.L., McCrone, C.J., Walter, B.J., Pratt K.B., Greenberg D.P. 2013. Automated Translation and Thermal Zoning of Digital Building Models for Energy Analysis. Proceedings of the IBPSA Building Simulation 2013 Chambery, 2013

Kämpf, J.H., Robinson, D., 2007. A simplified thermal model to support analysis of urban resource flows. Energy and Buildings 39, 445–453.

Kramer, R., van Schijndel, J., Schellen, H., 2012. Simplified thermal and hygric building models: A literature review. Frontiers of Architectural Research 1, 318–325.

Lehmann, B., Gyalistras, D., Gwerder, M., Wirth, K., Carl, S., 2013. Intermediate complexity model for Model Predictive Control of Integrated Room Automation. Energy and Buildings 58, 250–262.

Lepore, R., Renotte, C., Frère, M., Dumont, E., 2013. Energy consumption reduction in office buildings using model-based predictive control, in: Building Simulation 2013. pp. 2458–2464.

MathWorks Matlab. http://www.mathworks.de/products/matlab/

Maxima, a Computer Algebra System. http://maxima.sourceforge.net/

Meyer zu Eissen, S., Stein, B., 2006. Realization of Web-based simulation services. Computers in Industry 57, 261–271.

Miller, C., 2013. Automation of Common Building Energy Simulation Workflows Using Python [WWW Document]. URL

http://www.academia.edu/3410150/Automation_of_Common_Building_E
nergy_Simulation_Workflows_Using_Python (accessed 11.3.13).

Neto, A.H., Fiorelli, F.A.S., 2008. Comparison between detailed model simulation and artificial neural network for forecasting building energy consumption. Energy and Buildings 40, 2169–2176.

Nielsen, T.R., 2005. Simple tool to evaluate energy demand and indoor environment in the early stages of building design. Solar Energy 78, 73–83.

Oldewurtel, F., Parisio, A., Jones, C.N., Gyalistras, D., Gwerder, M., Stauch, V., Lehmann, B., Morari, M., 2012. Use of model predictive control and weather forecasts for energy efficient building climate control. Energy and Buildings 45, 15–27.

ÖNORM B 8110-3 2012. Teil 3, Vermeidung sommerlicher Überwärmung, Österreichisches Normungsinstitut (ON)

Pal, S., Roy, B., Neogi, S., 2009. Heat transfer modelling on windows and glazing under the exposure of solar radiation. Energy and Buildings 41, 654–661.

Peng, C., Wu, Z., 2008. Thermoelectricity analogy method for computing the periodic heat transfer in external building envelopes. Applied Energy 85, 735–754

Perez, R., Ineichen, P., Seals, R., Michalsky, J., Stewart, R., 1990. Modeling daylight availability and irradiance components from direct and global irradiance. Solar energy 44, 271–289.

Python Programming Language – Official Website, http://www.python.org/

Rumianowski, P., Brau, J., Roux, J.J., 1989. An adapted model for simulation of the interaction between a wall and the building heating system, in: Thermal Performance of the Exterior Envelopes of Buildings IV Conference, Orlando, USA.

See, R., Haves, P., Sreekanthan, P., O'Donnell, J., Basarkar, M., Settlemyre, K., 2011. Development of a user interface for the Energyplus whole building energy simulation program, in: Proceedings of Building Simulation. pp. 2919–2926.

Široký, J., Oldewurtel, F., Cigler, J., Prívara, S., 2011. Experimental analysis of model predictive control for an energy efficient building heating system. Applied Energy 88, 3079–3087.

VDI, 2012. Verein deutscher Ingenieure, Calculation of transient thermal response of rooms and buildings, Modelling of solar radiation, VDI 6007 / 3, 2012

Walton, G.N., 1979. The application of homogeneous coordinates to shadowing calculations. ASHRAE Transactions, 84 (1), 174–180

Wang, H., Chen, Q., 2013 Human-behavior orientated control strategies for natural ventilation in office buildings.

Winkelmann, F.C., 2001. Modeling windows in EnergyPlus. Proc. IBPSA, Building Simulation.

Weiler, K., Atherton, P., 1977. Hidden surface removal using polygon area sorting, in: ACM SIGGRAPH Computer Graphics. pp. 214–222.

Yahiaoui, A., Hensen, J., Soethout, L., Van Paassen, A.H.C., 2008. Developing web-services for distributed control and building performance simulation using run-time coupling, in: Proceedings of the 10th IBPSA Building Simulation Conference.

Zhao, H., Magoulès, F., 2012. A review on the prediction of building energy consumption. Renewable and Sustainable Energy Reviews 16, 3586–3592.

Zhou, Y., Liepe, J., Sheng, X., Stumpf, M.P.H., Barnes, C., 2011. GPU accelerated biochemical network simulation. Bioinformatics 27, 874–876.

7.2 Figures

Figure 1: Graphical presentation of the organization of the thesis at hand. 6

Figure 2: In the context of the thesis the model represents a central part. 7

Figure 3: Presentation of a zone to which an exterior wall is attached. 8

Figure 4: Presentation of a two-zone model to which an exterior wall is attached. The cut on the right hand side indicates that the model could be infinitely continued with further construction elements and zones. 11

Figure 5: RC presentation of a non-symmetric internal wall. Both surface nodes are coupled with the same zone. 12

Figure 6: RC presentation of a symmetric internal wall. Since the values for the thermal capacity and resistance are the same on both surfaces, one node can be removed. 12

Figure 7: According to the Weiler-Artherton algorithm the parts outside of the projection of the foremost polygon are clipped. Only the inside parts are sorted. (Gladt and Bednar, 2013) 19

Figure 8: Polygon clipping is applied after the relevant faces have been projected unto a plane orthogonal to the sun. (Gladt and Bednar, 2013) 19

Figure 9: Schematic representation of the objects passed to and returned from the interface of the implementation.0 21

Figure 10: In the context of the thesis the implementation concerns the model and the solver. 23

Figure 11: Simplified UML chart of the implementation of the RC model. 27

Figure 12: Schematic representation of processes distributed on different machines. ... 31

Figure 13: The original CAD model has to be transformed into a model that can be processed by the solver. 33

Figure 14: Carlsaw's approach is a function of time and penetration depth. 34

Figure 15: The solver represents the component that the model is passed for solving the relevant equation system. 36

Figure 16: The load Q_k occurs between two points that the solution is calculated for. Thus it is not taken into account by the solver. 38

Figure 17: Euler's method extrapolates the deviation in order to calculate y_{i+1}. 40

Figure 18: There is a linear dependency between the step size and the accuracy of results (represented by the average temperature deviation) with an implementation of the Euler method. 41

Figure 19: The linear correlation between accuracy and average step-size can only partly be observed with the LSODA implementation. 42

Figure 20: There is a linear correlation also between the accuracy of the energy calculations and the step size. 43

Figure 21: Speed of the LSODA solver compared with Euler solver and BDF solver respectively. Euler performs quite well as long as the required accuracy is low and time steps can be raised. LSODA outperforms Euler and BDF at almost any required accuracy. 44

Figure 22: Control systems rather represent interactions between individual components than a particular component themselves. 46

Figure 23: A temperature-controlled ventilation system needs to limit ventilation periods if the outside temperature causes a sharp decrease of the room temperature. 49

Figure 24: Schematic representation of typical dissipated energy and air exchange rates of natural and mechanical ventilation in summer. 50

Figure 25: Typical temperature profile and the corresponding power profile of the heating for a model where a thermostat-controlled heating is simulated 52

Figure 26: Typical temperature profile and the power profile of the heating for a model where the ideal heating power is continuously calculated 53

Figure 27: Dimensions of the model used with BESTEST MZ320 (Alexander et al., 2008) 55

Figure 28: Temperature profile and relevant cooling power profile of zone C according to BESTEST MZ320 (Alexander et al., 2008) in case of thermostat-based control 56

Figure 29: Temperature profile and relevant cooling power profile of zone C according to BESTEST MZ320 (Alexander et al., 2008) in case of an ideal cooling configuration 57

Figure 30: The reduction algorithm can remain independent of the implementation of the model and the solver. 60

Figure 31: If walls can be merged because the relevant physical properties and excitations are identical equations can be saved. 64

Figure 32: Zones are divided into groups. Zones are merged if their temperature profiles are similar. 68

Figure 33: Neighboring zones are left unmerged in order to achieve similar temperature profiles with the full-scale model and the reduced one. 70

Figure 34: Standard floor of an office building with an unlimited number of identical floors above and below. 72

Figure 35: The temperature profile of the zone marked green is of interest. Applying the reduction algorithm according to 4.7 a model is created where only the colored zones are left unmerged. 73

Figure 36: Annual temperature profiles of the zone marked green in Figure 35 in the context of the full-scale model and of the reduced model. 74

Figure 37: Short term temperature profiles of the zone marked green in Figure 35 in the context of the full-scale model and of the reduced model. 74

Figure 38: Standard floor of an office building where the zones representing office rooms (blue) are merged. 76

Figure 39: Annual temperature profiles of the corridor within the full-scale model and the model subjected to zone grouping according to Figure 38. 77

Figure 40: Short term temperature profiles of the corridor within the full-scale model and the model subjected to zone grouping according to Figure 38. 77

Figure 41: Short term profiles of the overall cooling power matching the temperature profile in Figure 40. 78

Figure 42: Annual temperature profile of the corridor within the full-scale model and the one subjected to zone grouping according to Figure 38. 79

Figure 43: Three days' temperature profiles of the corridor in the context of the full-scale model and of the reduced model according to Figure 38. 80

Figure 44: The architecture of simulation system is influenced by the decision about whether a client server or a standalone implementation is accomplished. 83

Figure 45: Overview of an architecture with three optional layers. Missing layers have to be replaced by the client PC. 84

Figure 46: Overview of a synchronous request, response cycle. There are no middle layers in this setup 85

Figure 47: Overview of a setup including a layer between the client and the simulation servers. The client disposes of a callable function or web service itself. 86

Figure 48: Overview of a setup including a layer between the client and the simulation servers. The client does not dispose of a callable function or web service itself. Instead it has to poll for the result. 87

Figure 49: With the help of a plugin MS Excel can be used as a web service client. The code for accessing the web service is written in MS Visual Studio. A representation of the model is held in several spreadsheets of an Excel file. 93

7.3   Tables

Table 1: Comparison of C++, Matlab and Python in terms of qualification for building simulation 26

Table 2: Specifics of the individual objects used in the implementation of the RC model 30

Table 3: Material properties of the walls used with BESTEST MZ320 (Alexander et al., 2008) 55

Table 4: Summary of different control strategies. 59

Table 5: Summary of reduction methods 64

Table 6: Example for wall merging according to Figure 31 65

Table 7: Particulars of the individual construction elements 71

Table 8: Particulars of the zones.. 71

Table 9: Particulars of the windows 71

Table 10: Summary of results of a "free-float" simulation of a full-scale model compared with a reduced on according to Figure 35. 75

Table 11: Summary of overall results of a full-scale model compared with a reduced on according to Figure 36. 78

Table 12: Summary of results for the corridor in the context of a model reduced according to Figure 35. Windows are openable in summer and heating system is set into ideal operating mode in winter. 80

8 Appendix

A FULLY AUTOMATED CALCULATION OF SHADOW CASTING WITH MATRIX-BASED COORDINATE TRANSFORMATIONS AND POLYGON CLIPPING

Matthias Gladt[1] and Thomas Bednar[1]
[1]Vienna University of Technology, Vienna Austria

ABSTRACT

Short wave radiation has a significant influence on a building's thermal behavior. The calculation of the projected sunlit surface fraction is therefore an indispensable prerequisite for running a meaningful simulation. It consists of at least two major steps. One is to solve the visibility problem of all surfaces of the building that is investigated. This means to set up an order in which one surface may cast a shadow on another.

The other step is to calculate the projected sunlit surface fraction based on the order obtained in the first step.

There are basically two technologies to address both problems. One is based on OpenGL and the other one on polygon clipping algorithms. The paper at hand will outline a polygon-based implementation and give the pros and cons compared with an OpenGL-based implementation.

INTRODUCTION

The visibility problem as well as the need for the calculation of the projected sunlit surface fraction (PSSF) usually occurs in the course of a building simulation where short wave radiation shall be taken into account. Both issues have been addressed in papers as early as in the 1970s. Walton (1979) compares two algorithms of calculating the shadow. One, the "discrete element analysis" is the basis for modern GPU-based calculations, the other, "overlapping polygons", focuses on finding a more accurate solution for the calculation of the PSSF. Newell, Newell and Sancha outline a method for solving the hidden surface problem (1972) which is further developed by Weiler and Artherton (1978). Artherton, Weiler and Greenberg (1977) explain how these methods can be integrated with computer-aided design (CAD) rendering at the time. Grau and Johnson (1995) show how Weiler and Artherton's polygon clipping algorithm (1978) can be implemented.

Even with modern computers polygon clipping remains the most consuming part of the calculation if accurate results are required. Therefore efficient polygon clipping algorithms are vital for a fast computation. Vatti (1992) presents a polygon clipping algorithm on which some modern implementations are based (e.g. Murta, 2009). Although Vatti's algorithm can still be slightly improved (Greiner and Hormann, 1998) the consumption of CPU time grows with the average number of intersections of polygons on the order of $O(nm)$ where n and m denote the numbers of edges of the polygons to be intersected.

The calculation of the PSSF consists of several steps. Most of them have already been dealt with in literature but hardly any publications exist outlining the whole process from providing an interface for the input data over processing all steps needed to calculate the PSSF to returning the results in a usable form.

This paper will outline in detail how this goal can be achieved by the means of a modern programming language in general and by the example of the Python programming language in particular.

It will also show how a new implementation can be based on some sophisticated solutions for the individual steps to calculate the PSSF and how an easy integration with existing building simulations can be achieved.

Furthermore it will point out where alternatives to the solution provided for a particular step of the calculation exist and it will elaborate on the pros and cons of these alternatives.

POLYGON-BASED APPROACH

State of the art

Jones, Greenberg and Pratt (2011) define the projected sunlit surface fraction (PSSF) as

$$I_B \left(A_S / A_T \right) \cos\theta \qquad (1)$$

where I_B is the intensity of radiation (i.e. the sunlight), A_S the sunlit surface area, A_T the total surface area and θ the angle of incidence of sun's rays with respect to the surface's normal vector. A_S on its part is defined as

$$A_S = A_T - A_D \qquad (2)$$

where A_D (D...dark) is the part of A_T that a shadow is cast on.

A_D is the factor to be calculated with the help of the implementation at hand. Since the sunlight is in the focus of interest rather than one or more spot lights all rays can be assumed parallel.

A sample scene (Figure 1), shall be used to illustrate the following approach. There are two faces, one casting a shadow on the other. Although the example at hand is a simple one all considerations are equally valid for more complex scenarios.

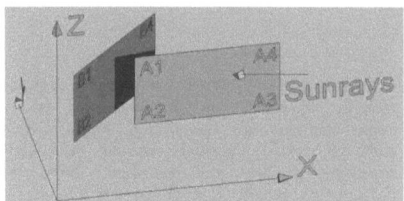

Figure 1 Scene with two surfaces, one casting a shadow on the other

Shadow mapping

Williams (1978) proposes to render the scene from the position of the light and to generate a map with depth values representing the distance of a particular object from the source of light. The values of the so called "depth map" are updated whenever an object is added to the scene whose distance from the source of light is smaller than the current value from the depth map. In the end the depth map holds the values representing the distance(s) of the foremost object(s) seen from the position of the light.

We adopt the first part of Williams' idea to render the scene from the light's point of view (Figure 2) but we modify his approach in which he stores a complete depth map.

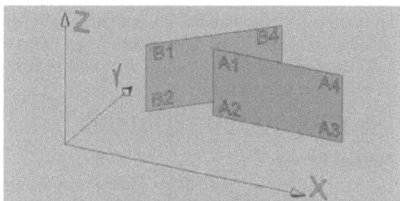

Figure 2 Same scene from light's point of view. Rendering the scene from the light's point of view provides all faces exposed to direct sunlight whereas any other faces, not visible from this point, remain in the shadow

Instead we will only compare depth values at some points and use it as a basis for the depth sort of all objects.

Weiler's and Artherton's hidden surface removal

Weiler and Artherton (1978) present a method to set up an order based upon the depth values of the individual polygons where the polygons represent the enveloping surfaces of an object. For the sake of simplification only flat faces were used in Figure 3 which can also be thought of as detached slim walls.

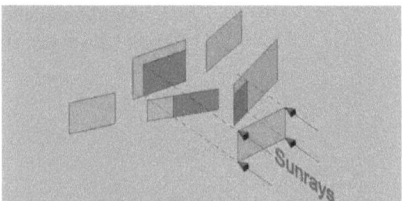

Figure 3 Weiler's and Artherton's algorithm of solving the visibility problem

Weiler and Artherton apply their sorting method only to polygons that are partly or completely covered by the foremost polygon seen from the position of the light. They store these polygons in an "inside list". The inside list has to meet the criterion that none of its polygons must be in front of the polygon that was assumed foremost. If it cannot be met a new inside list has to be set up recursively with the polygon violating the above criterion as the new foremost.

Polygons which are not part of the inside list as well as the parts of the polygons from the inside list that are not covered by the currently foremost polygon are temporarily stored in an "outside list". The outside list will become the new inside list after the original one has been finished processing.

So Weiler and Artherton's algorithm requires that two geometrical problems are solved:

- Determine whether two polygons overlap from the point of view of the light.
- Determine which of two polygons is more to the front.

And both those problems require that

- a 3D scene is projected onto a 2D projection plane (Figure 4) and that
- the 2D projection plane is transformed from the original 3D coordinate system into a 2D coordinate system in order to apply common polygon clipping algorithms.

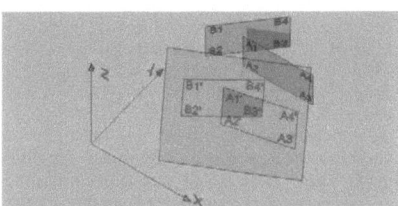

Figure 4 Projection of a 3D scene on a plane orthogonal to the sun. The overlapping part in the projection plane represents the shadow being cast from one polygon onto the other. However, without solving the visibility problem no conclusions can be drawn as to whether polygon A casts a shadow on polygon B or vice versa.

Transformation from 3D into 2D

The sequence in which the two steps are executed is arbitrary. The relevant rotation matrix and projection matrix are not the same in both cases, though. We choose to rotate our coordinate system before and apply the projection afterwards.

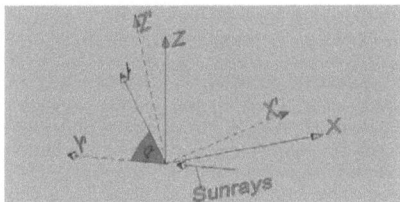

Figure 5 Rotation of the coordinate system to have one axe parallel to the sunrays

The choice of the axis to be aligned with the direction of the light is of no relevance for the transformation into a 2D coordinate system. It can be X as well as Y or Z. In the example illustrated in Figure 5 we chose Y.

The rotation matrix in order to align Y with the direction of the light is set up as

$$\tilde{M}_{rot} = \begin{bmatrix} r_x^2(1-\cos\alpha)+\cos\alpha & r_xr_y(1-\cos\alpha)-r_z\sin\alpha & r_xr_z(1-\cos\alpha)+r_y\sin\alpha \\ r_xr_y(1-\cos\alpha)+r_z\sin\alpha & r_y^2(1-\cos\alpha)+\cos\alpha & r_yr_z(1-\cos\alpha)-r_x\sin\alpha \\ r_xr_z(1-\cos\alpha)-r_y\sin\alpha & r_yr_z(1-\cos\alpha)+r_x\sin\alpha & r_z^2(1-\cos\alpha)+\cos\alpha \end{bmatrix} \quad (3)$$

with $\bar{r}$ as the rotational axis through the origin and α the rotation angle.

$$\bar{r} = \begin{bmatrix} r_x \\ r_y \\ r_z \end{bmatrix} \quad (4)$$

$\bar{r}$ is orthogonal to the Y-axis and the direction of the light $\bar{s}$:

$$\bar{r} = \begin{bmatrix} 0 \\ 1 \\ 0 \end{bmatrix} \times \begin{bmatrix} s_x \\ s_y \\ s_z \end{bmatrix} \quad (5)$$

and

$$\cos\alpha = s_y \quad (6)$$

The rotation of a single point can be calculated via the matrix multiplication of $\tilde{M}$ with the relevant coordinates:

$$\begin{bmatrix} p_x' \\ p_y' \\ p_z' \end{bmatrix} = \tilde{M} \begin{bmatrix} p_x \\ p_y \\ p_z \end{bmatrix} \quad (7)$$

Calculating the rotation of all n vertices with a single matrix multiplication supported by Python's NumPy results to:

$$\begin{bmatrix} p_{1,x}' & p_{2,x}' & \cdots & p_{n,x}' \\ p_{1,y}' & p_{2,y}' & \cdots & p_{n,y}' \\ p_{1,z}' & p_{2,z}' & \cdots & p_{n,z}' \end{bmatrix} = \tilde{M}_{rot} \begin{bmatrix} p_{1,x} & p_{2,x} & \cdots & p_{n,x} \\ p_{1,y} & p_{2,y} & \cdots & p_{n,y} \\ p_{1,z} & p_{2,z} & \cdots & p_{n,z} \end{bmatrix} \quad (8)$$

The new y'-coordinate of a vertex represents its distance from the position of the light reduced by the distance of the origin from the position of the light. Since the second term is constant it can be ignored in the following considerations.

The plane receiving a parallel projection is represented by the equation:

$$p_x + p_y + p_z = 0 \quad (9)$$

The direction of the projection can be represented by the vector:

$$\bar{d} = \begin{bmatrix} d_x \\ d_y \\ d_z \end{bmatrix} \quad (10)$$

Thus the matrix for the parallel projection results to:

$$\tilde{M}_{pq} = \begin{bmatrix} (d_yp_y+d_zp_z)/c & -d_xp_y/c & -d_xp_z/c \\ -d_yp_x/c & (d_xp_x+d_zp_z)/c & -d_yp_z/c \\ -d_zp_x/c & -d_zp_y/c & (d_xp_x+d_yp_y)/c \end{bmatrix} \quad (11)$$

where

$$c = d_xp_x + d_yp_y + d_zp_z \quad (12)$$

If the direction of projection is parallel to the Y'-axis:

$$\bar{d} = \begin{bmatrix} 0 \\ 1 \\ 0 \end{bmatrix} \quad (13)$$

Then the projection matrix obviously defaults to:

$$\tilde{M}_{proj} = \begin{bmatrix} 1 & 0 & 0 \\ 0 & 0 & 0 \\ 0 & 0 & 1 \end{bmatrix} \quad (14)$$

So the y'-values of all vertices can be omitted in order to apply a 2D polygon clipping library and to calculate the overlap of the projections of the individual polygons.

The order in which the polygons are overlapped is given by the inside list mentioned above. Nevertheless the order of the inside list has to be checked because there is no guarantee that the original order of the inside list is correct. (Figure 6)

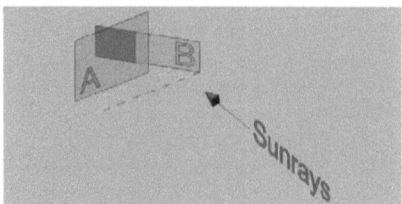

Figure 6 Scenario causing an error in the depth sort based on the lowest y' – coordinate of each polygon. The inside list needs to be re-created recursively.

The check can be made at any point within the overlapping part of the two polygons in the projection plane and consists of comparing the relevant y' – values of the two polygons. If the original order turns out to be erroneous then a new inside list has to be created recursively.

Calculation of shadowed parts

The shadowed part of a particular polygon can easily be calculated from the area of all clipped parts of the projections resulting from the other polygons further at front:

$$A_D \cos\theta = \sum A_{ClipProjPoly} \quad (15)$$

The total area can be obtained from the original projection of a particular polygon.

$$A_T \cos\theta = A_{ProjPoly} \quad (16)$$

The polygon clipping library that was used to figure out the clipped parts of a polygon's projection was Murta's general polygon clipper library (GPC). It is based on Vatti's generic solution to polygon clipping (1992) and can also process convex polygons and polygons with holes which is needed to take windows into account.

Rädler provides a python binding for the GPC and binds clipping operations to the standard operators like +, -, & etc.

OPENGL-BASED ALTERNATIVE

An alternative way of calculating shadows and the PSSF respectively is based on OpenGL. (Jones, Greenberg and Pratt, 2011) OpenGL is intended for accessing the graphics processing unit (GPU) of a computer in order to manipulate the picture sent from the computer to the screen and to make use of the parallel computing potential of the GPU as efficiently and as directly as possible. The idea of exploiting GPUs for scientific calculations that can be parallelized has come up only during the recent years. However, the GPU-centric design of OpenGL as well as its original purpose makes it often difficult to port scientific implementations to OpenGL.

Although shadow calculation appears to be an explicitly graphic problem also seemingly peripheral tasks such as figuring out the coordinates of a shadow polygon from the signal of the GPU or counting pixels representing the shadowed part of a model can become troublesome.

Another problem with OpenGL is related to the diversity of GPUs and to the unreliable compatibility of GPU drivers with current programming specifications. Where some manufactures comply well with agreed standards others don't. The OpenGL wiki states that a recent driver is needed for the compatibility with modern OpenGL versions and recommends to also try out beta versions of drivers in order to get access to bugfixes.

Therefore problems related to the driver for the GPU which is often needed in a particular version may occur as well as problems with the installation of the programming framework needed to develop OpenGL applications. Prior to the attempt to implement a complete shadow mapping calculation with OpenGL the installation of the OpenGL utility toolkit (GLUT) had to be completed. The installation process was also troubled by inconsistencies of particular versions which produced error messages of little value.

Baker (2009) runs a website on which he not only explains the theory of a GPU-based implementation of shadow mapping but also gives a concrete example in C++ programming language. His example was modified to create a scene representing the facades of a virtual assembly of buildings.

Drawing objects in OpenGL can be done with the OpenGL utility toolkit (GLUT). GLUT disposes of a few functions to draw simple geometric forms, such as triangles, quads or polygons. These functions can be used to create also complex structures such as a building envelope.

Rendering a scene with shadows requires the relevant transformation matrices in OpenGL. A transformation matrix can be obtained via the glGetFloat command after the desired operations have been applied to the matrix currently on top of the OpenGL matrix stack. The object that is intended to hold the relevant matrix is passed as an argument. There are two types of matrices which are needed to implement shadow mapping in OpenGL:

- Matrices defining a perspective projection from the position of the light and from the position of the camera respectively.
- Matrices defining a view transformation to render the scene from a particular point, in this case the position of the light and the position of the camera respectively.

OpenGL-based shadow mapping builds upon Williams' approach (1978): A scene needs to be rendered three times in order to implement shadow mapping in OpenGL. In the first pass the scene is drawn from the position of the light. This is done by applying the light-projection matrix in the GL_PROJECTION mode of OpenGL and by applying the light-view matrix in the GL_MODELVIEW mode. The depth values of all objects within the current field of view are stored in a shadow map texture which will be used in a successive step to identify the lit-up parts of the scene.

In the second pass the whole scene is rendered in dim light from the position of the camera by applying the camera-projection matrix and the camera-view matrix and setting the required properties to set up dim light.

The third pass is where the depth map comparison is set up. Therefore the depth values from the shadow map texture of the first pass are compared with the depth values of a newly generated texture representing the objects of the scene in sunlight. OpenGL provides a function (*glAlphaFunc*) that can be used to compare the depth values of two textures and to discard all fragments of the texture disposing of a higher depth value than the shadow map texture. This way only those parts of the newly generated texture are rendered which are in direct light.

Figure 7 Image of correct OpenGL-based calculation of shadow casting

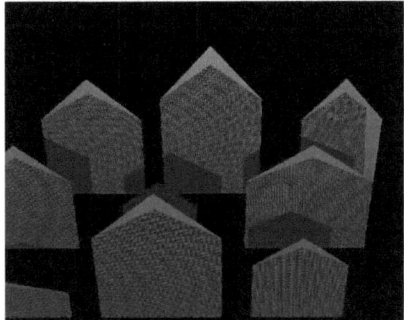

Figure 8 Image of OpenGL-based calculation of shadow casting with unwanted artifacts

The attempt to develop an OpenGL based shadow mapping implementation was partly successful (Figure 7) but partly the results rendered had notable artifacts (Figure 8) which might be due to the fact that the quality of the results of different GPUs may vary, depending on the manufacturer and the model. The GPU used for the attempts at hand was a rather simple onboard one.

Rendering a scene with shadows is usually not enough for determining the PSSF for each surface of interest. Therefore the number of pixels has to be determined. Jones et al. (2011) present a method based on Yezioro's and Shaviv's approach (1994) where they render a particular scene or a portion thereof to two image buffers, once with shading devices and once without. Then the number of pixels held in the two image buffers is tallied, and the fraction of each surface which is shadowed can be calculated. Bartz et al. present a method of OpenGL-assisted occlusion culling for large polygonal models where they use a stencil buffer to run an occlusion test.

Since 2003 OpenGL provides an extension, the ARB_occlusion_query, whose functionality is demonstrated by Harris (2005). Occlusion queries can be used to track the number of fragments or samples that pass the depth test. For this purpose all relevant objects have to be wrapped into an occlusion query by calling *glBeginQueryARB* and *glEndQueryARB* respectively. The command *glGetQueryObjectuivARB* writes the number of pixels of the relevant object or texture into a variable which is passed as an argument. A pixel count of the texture representing the parts that are in direct light and of the whole scene can be used to determine the PSSF.

EVALUATION

The polygon-based implementation

The implementations were tested with a notebook with an Intel P6100 CPU with 2.0 GHz. The polygon-based implementation was applied to an arbitrary number of randomly generated polygons regarding their size and orientation. The polygons were placed in rows of five with the rows at a randomly generated distance from each other. Intersections between polygons were avoided. The light source was assumed horizontal in the direction of where the rows of polygons were aligned. Each polygon had four vertices and three windows of different size.

Table 1 Calculation time for scenarios with an arbitrary number of randomly generated polygons

NUMBER OF POLYGONS	NUMBER OF VERTICES	CALCULATION TIME IN SECONDS
200	800	0.1
400	1,600	0.3
800	6,400	1.1
1,600	12,800	4.2
3,200	25,600	15.9
6,400	51,200	67.5
9,600	76,800	151.6
12,800	102,400	264.7

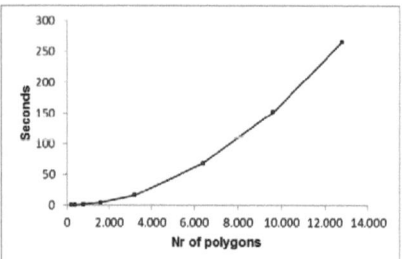

Figure 9 Performance of GPC-based implementation. Processing time rises in a ratio of the power of two of the number of input polygons

Comparing OpenGL with polygon clipping

The OpenGL-based implementation as well as the polygon-based one has been tested with a scenario where the shadow that a treetop casts on a flat wall is calculated. The leaves of the treetop were modeled as triangles whose edge lengths were reduced between the individual simulation runs in order to raise the number of leaves.

Figure 10 Image of OpenGL-based simulation of a treetop casting a shadow on a flat wall.

With the OpenGL implementation the number of leaves was set to about 500 at the first simulation run. It was doubled in both directions after each run ending up at about 500.000 leaves after five runs. The shadow map buffer was set to a size of 768^2. There was practically no delay between the start of the simulation run and the rendering of the scene to the screen. However already at the third simulation run (8.000 leaves) it was not possible to make out a meaningful result with the naked eye. With the chosen shadow map buffer about 74 pixel can be allocated per leaf presumed the leaves do not cover each other. So no exact result can be expected if the number of details comes close to the size of the shadow map buffer.

Jones et al. (2011) propose to set the viewport to the part (e.g. a surface) of a relevant scenario that the PSSF shall be calculated for. Thus the accuracy of the result can be augmented.

Whereas in general the exactness of the result suffers and the performance is excellent with OpenGL for scenarios containing a high number of details the opposite is the case with implementations based on polygon clipping. The scenario from Figure 10 was also calculated with the polygon-based implementation. A simulation run with 2.500 leaves took about 45 seconds, one with 10.000 leaves about 550 seconds. The difference from the figures in Table 1, illustrated in Figure 9 is due to differences of the geometric circumstances of the test scenarios. In the first test scenario some polygons were covering others whereas with the treetop simulation all leaves were considered to be at the same depth. The result for the treetop simulation included not only the exact PSSF for the wall in the background but theoretically also the PSSF for every single leaf.

Accuracy

The accuracy of the results obtained by an OpenGL calculation depends on the size of the shadow map buffer and the geometry of the objects within the relevant viewport. Jones et al. (2011) show that the error for the calculation of a single surface is usually negligible. However, depending on the geometry of the scene and the position of the viewport, more than one surface needs to be taken into account in a single calculation run. This is when the calculation error may rise depending on the number of obstructing objects in the scene.

Several runs with different viewports in order to cover the whole scene may be required to obtain good accuracy.

The error of a polygon-based calculation depends on the accuracy of the intersections of the projections of the individual edges of two surfaces which is usually also negligible.

CONCLUSION

The calculation of shadowed areas is indispensable for a running comprehensive building simulation. There are substantially two different types of calculations: Polygon-based ones and GPU-based ones.

The paper at hand demonstrates how a complete polygon-based implementation can be set up, starting with a list of coordinates representing the envelopes of buildings and ending with the calculation of the PSSF. Shadow calculation is theoretically often reduced to clipping polygons. Many publications ignore that the prerequisites for applying a polygon clipping algorithm need to be fulfilled before. These prerequisites comprise the projection of 3D objects onto a 2D projection plane as well as solving the visibility problem. Transformations of 3D forms into a 2D coordinate system require the relevant matrices to be set up and the appropriate programming tools or languages to perform the associated mathematical operations. A matrix-based transformation from 3D into 2D implies a significant performance gain compared to vector-based calculations of intersections for all points.

Python is one well known programming language written in C that disposes of sophisticated libraries such as NumPy supporting these operations. Also some polygon clipping libraries, such as the GPC, dispose of a python binding which eases a homogeneous implementation regarding the programming language. Weiler and Artherton do not only present a polygon clipping algorithm in their paper (1977) but also a way to solve the hidden surface problem. Whereas the hidden surface removal algorithm is not critical regarding the overall performance of the implementation a polygon clipping library can be. Jörg Rädler's binding of the GPC was used in the implementation at hand. The GPC itself is based on Vatti's solution to polygon clipping. It supports concave polygons (which is not the case with Weiler's and Artherton's algorithm) and holes which is needed in order to take windows into account.

GPU-based approaches of shadow calculation are usually based on shadow mapping. The installation and programming effort is bigger compared with the polygon-clipping-approach. However its performance can well be used in real time calculations and with complicated structures because the approach makes use of the available GPU the best possible way. Since GPU-based implementations return a result per pixel the accuracy of the calculation depends on the size of the chosen viewport. Jones et al. (2011) have shown that with a viewport set in a sensible way the solution is accurate enough for most requirements.

With complex scenarios the full advantage of GPU driven calculations can be used most efficiently.

The effort of setting up the programming environment as well as preparations for rendering the scene such as building up the model in OpenGL and setting the right field of view, etc. is substantial.

Thus based on the current state of programming technology and scientific knowledge the authors of the paper at hand favor the polygon-based approach to calculate shadowing.

Acknowledgements

Research funded by the Zentrum für Innovation und Technologie GmbH (ZIT) and the Federal Ministry for Transport, Innovation and Technology.

The authors would also like to thank Nathaniel Jones for providing some publications relevant for the this paper.

REFERENCES

Artherton P., Weiler K., Greenberg D., 1978. Polygon shadow generation. Proceedings of the 5th annual conference on Computer graphics and interactive techniques, p.275-281, August 23-25.

Baker P., 2009. Shadow Mapping [online]. Available from http://www.paulsprojects.net/tutorials/smt/smt.html [Accessed 4 November 2012]

Bartz D., Meißner M., Hüttner T., 1999. OpenGL-assisted occlusion culling for large polygonal models, Computer & Graphics, vol. 23, no. 5, pp. 667-679

Grau, K. and Johnsen, K., 1995. General shading model for solar building design. ASHRAE Transactions, 101 (2), 1298–1310.

Greiner G., Hormann K., 1998 Efficient clipping of arbitrary polygons, ACM Transactions on Graphics, 17 (2), pp. 71–83

Harris K., 2005. Occlusion Query [online]. Available from http://www.codesampler.com/oglsrc/oglsrc_7.htm [Accessed 4 November 2012]

Jones N. L., Greenberg D. P. and Pratt K. B., 2011. Fast computer graphics techniques for calculating direct solar radiation on complex building surfaces. Journal of Building

Khronos Group, 2010. OpenGL, Open Graphics Library [online]. Available from http://www.opengl.org/ [Accessed 4 November 2012]

Khronos Group, 2010. GLUT, The OpenGL Utility Toolkit [online] http://www.opengl.org/resources/libraries/glut/ [Accessed 4 November 2012]

Khronos Group, 2010. ARB_occlusion_query [online] http://www.opengl.org/registry/specs/ARB/occlusion_query.txt [Accessed 4 November 2012]

Murta, A., 2009. General polygon clipper library (GPC) [online]. Available from: http://www.cs.man.ac.uk/~toby/gpc/ [Accessed 4 November 2012].

Newell M. E., Newell R. G., Sancha T. L., 1972. A solution to the hidden surface problem. Proceedings of the ACM annual conference, August 01-01, 1972, Boston, Massachusetts, United States

NumPy, package for scientific computing with Python [online] http://numpy.scipy.org/ [Accessed 4 November 2012]

Python Programming Language – Official Website, http://www.python.org/

Rädler J., Polygon [online]. Available from http://www.j-raedler.de/projects/polygon/ [Accessed 4 November 2012]

Vatti, B.R., 1992. A generic solution to polygon clipping. Communications of the ACM, 35 (7), 56–63.

Walton, G.N., 1979. The application of homogeneous coordinates to shadowing calculations. ASHRAE Transactions, 84 (1), 174–180

Weiler K., Artherton P., 1977. Hidden surface removal using polygon area sorting, ACM SIGGRAPH Computer Graphics, 11 (2), 214–222

Williams L., 1978. Casting curved shadows on curved surfaces, Proceedings of the 5th annual conference on Computer graphics and interactive techniques, p.270-274, August 23-25, 1978

Yezioro, A. and Shaviv, E., 1994. Shading: a design tool for analyzing mutual shading between buildings. Solar Energy, 52 (1), 27–37

A NEW METHOD FOR THE CALCULATION OF THE SKY VIEW FACTOR FOR NON-RECTANGULAR SURROUNDINGS

Matthias Gladt[1] and Thomas Bednar[1]
[1]Vienna University of Technology, Vienna Austria

ABSTRACT

The sky view factor (SVF) represents the relationship between the visible portion of the sky and the portion covered by the surrounding objects.

It is significant e.g. in urban zones where the nightly cooling of a building has an intense correlation with the long-wave radiation emitted by the surrounding objects.

The methods of calculating the SVF comprise fisheye-lens photographs analysis as well as mathematical models or image processing. A well-known method for the calculation is based on a vector-based presentation of the model. The projection of the model onto the hemisphere is divided into slices to approximate the calculation of the resulting polygon with the calculation of the SVF for the rectangles resulting from the slices of the previous step.

The approach has two major drawbacks: The process of discretization slows the calculation down and the accuracy of the result depends on the particularity of discretization. This paper presents a method where no more discretization is needed which results in lower calculation times and more accurate results.

INTRODUCTION

Especially in urban areas the nightly cooling of the buildings is limited because the densely built-up area prevents an effective heat rejection as it takes place in rural areas. Buildings act as heat accumulators but whereas in rural areas they practically only give off heat, in urban areas they also receive heat from their neighboring buildings. Gál et al. (2007) coin the term, "urban heat island" to clarify the relevance of surrounding objects for the heat balance of a particular building. A common way of calculating the sky view factor (SVF) which is elaborated in many publications is based on graphical methods and in particular on hemispherical photographs with a fisheye photographic lens. (E.g. Watson and Johnson, 1988)

However Matuschek and Matzarakis (2010) point out that the process of taking a picture is error-prone and can usually only be completed if an existing environment shall be evaluated. So it may not be possible to run simulations in an early phase of a project where only computer models exist. They evaluate "SkyHelios", a software tool, which can be used to calculate the SVF. It's based on digital elevation models of the relevant scenario and returns promising results.

Rakovec and Zakšek (2012) take into account the diffuse tilt factor in addition to the SVF. They present a method for the calculation of the complete diffuse radiation but they restrain to a flat slope rather than to projections from arbitrary objects.

Bradley, Thornes and Chapman (2001) focus on calculating the SVF value from hemispherical photographs with a fisheye photographic lens. They suggest using GPS in connection with it to avoid errors when building up a SVF database. They also take into account the density of the urban development to make out differences in the SVF value for particular classes of urban regions.

Gál et al. compare the SVF values received by a fisheye-lens photograph analysis with values calculated from an urban vector database. Their method of calculating the SVF from a vector database is analyzed in detail on the following pages. The fisheye-lens photographs are divided into a number of concentric annuli of equal width, each representing an interval of zenith angles. The results of both methods deviate by an average value of 0.106 (regarding the relevant SVF values) from each other which is mainly owed to the fact that vegetation is not in the database used as input for the calculations.

Matzarakis and Matuschek (2007) present "Ray-Man", a model used for the calculation of short- and long-wave radiation fluxes on the human body. RayMan also provides an interface for free drawings and fish-eye photographs to calculate the SVF.

Johnson and Watson (1984) present a method where they set up a polar coordinate system to calculate SVF values for finite and infinite canyons enclosed by walls of constant height. The method outlined in what follows is partly related to their approach but is based on a different coordinate system. Besides also walls with varying height with respect to the longitudinal axis are considered.

Källblad (1999) presents a generic function to calculate the SVF value and Oke (1987) who defines the sky view factor (SVF) as the ratio of the amount of the sky "seen" from a given point on a surface to that potentially available also gives values for commonly occurring geometric arrangements like valleys or slopes or a basin of a constant radius and of constant height. Based on these values the SVF can also be used to quantify the characteristics of a building's surroundings which are significant for the amount of heat that is received.

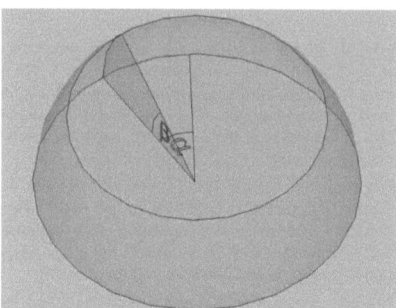

Figure 1 Basin with the opening angle of β representing the projection of a virtual surrounding of constant height on the hemisphere

The SVF for the whole basin (Figure 1) given by Oke is:

$$\psi_{sky} = \cos^2 \beta \qquad (1)$$

So the SVF in this case represents the area of the opening of the basin to the area of the middle circle quotient. If only the fraction represented by α shall be taken into account according to Gál et al. (2007) the SVF results to:

$$\psi_{sky} = \cos^2 \beta \left(\frac{\alpha}{2\pi} \right) \qquad (2)$$

The calculation methods presented in the paper at hand are partly based on Equation 1 and 2.

A solution for the calculation of the SVF for generic surroundings in terms of geometry is outlined. Gál et al. also base their approach of calculating the SVF from a vector database on Oke's value of the SVF for the basin mentioned above by dividing the hemisphere into slices and calculating the SVF for each slice. The drawbacks of their solution may regard performance issues and accuracy depending on the chosen level of discretization. With the raise of resolution of discretization not only the accuracy of the result may improve but also the amount of time consumed by a single simulation run may go up. The solution in the paper at hand avoids these drawbacks by presenting a new calculation method where no more discretization has to be made.

METHOD

Transformation of the scenery into a new coordinate system

For demonstration purposes we chose a simple scenario with two faces representing two of the enveloping surfaces of an object. They can also be interpreted e.g. as the facades of two buildings.

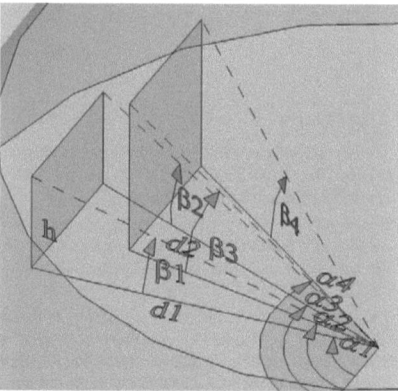

Figure 2 Two faces in relation to a thought part of the hemi-sphere. Vertical angles are labeled with β_{1-4}, horizontal ones with α_{1-4}.

The calculation of the SVF is based on the central projection of the relevant scenario. The origin is set at the point that the SVF shall be referred to. The projection of the surrounding buildings has to be specified by a function of the horizontal angle, α, returning the vertical angle β rather than by a set of Cartesian coordinates. The transformation from the Cartesian coordinate system into a new α, β coordinate system brings along a distortion of originally straight lines.

In the example of Figure 2 vertical edges remain vertical in the projection, horizontal edges, however, must be specified by the tangent of the height to distance quotient. (Figure 3)

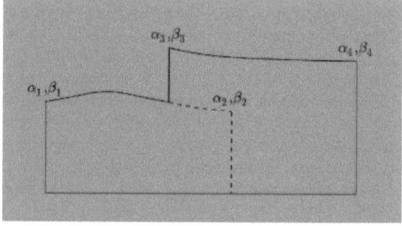

Figure 3 Schematic representation of the central projection of the same scenario from the position of the origin as a function of the horizontal angle α.

The exact shape of non-vertical edges depends on the geometric circumstances. In the example at hand the calculation of β_1 and β_2 is trivial:

$$\beta_1 = \arctan\left(\frac{h_1}{d_1}\right) \quad (3)$$

and

$$\beta_2 = \arctan\left(\frac{h_1 + \Delta h}{d_2}\right) \quad (4)$$

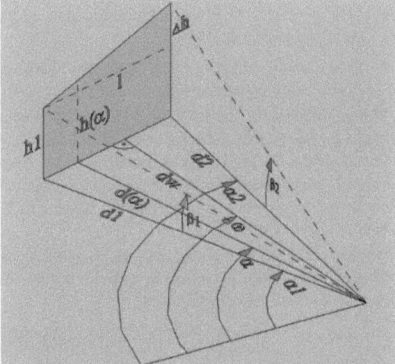

Figure 4 Geometry of one of the faces of the length, l, related to the point that the SVF refers to. d_w denotes the distance of the face from the origin and ω the angle of the line orthogonal to the face.

From Figure 4 it can be seen that the function describing β between β_1 and β_2 results to:

$$\beta(\alpha) = \arctan\left(\frac{h(\alpha)}{d(\alpha)}\right) \quad (5)$$

and

$$d(\alpha) = \frac{d_w}{\cos(\omega - \alpha)} \quad (6)$$

$$h(\alpha) = h_1 + \Delta h \cdot \frac{d_1 \sin(\omega - \alpha_1) - d(\alpha) \sin(\omega - \alpha)}{l} \quad (7)$$

So the originally horizontal straight edges of the two faces become curves in the projection illustrated in Figure 3.

Since only the difference between two angles is used in the calculations any reference axis can be chosen for all relevant angles.

Discretization approach

The projection resulting from the previous example is displayed in Figure 5.

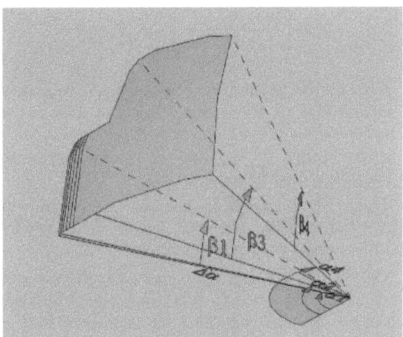

Figure 5: Resulting projection of two faces onto the hemisphere discretized by dividing it into slices of equal width.

Once the projection on the hemisphere as a function of the horizontal angle, α, is complete the SVF can be calculated based on Equation 2.

Figure 5 illustrates how Gál et al. (2007) divide the projection into slices, each covering the same fraction of the angle α, $\Delta\alpha$. They choose the middle points of the intervals as values for β. However under some unfavorable conditions the middle point may not present the average value of β. (Figure 6) A higher level of discretization can improve the accuracy of the result but will have a negative impact on the performance.

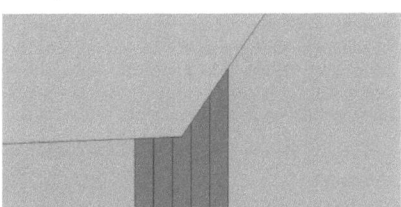

Figure 6: Discretization matches well left and right of the corner of the model. The slice in the middle produces an inaccurate result because its middle value is too low.

Functional based approach

An alternative approach is based directly on Equation 2. The SVF of a fraction of the basin can also be interpreted as the result of an integral:

$$\psi_{sky} = \cos^2\beta \frac{\alpha_2 - \alpha_1}{2\pi} = \frac{\alpha_2 - \alpha_1}{2\pi} - \frac{1}{2\pi}\int_{\alpha_1}^{\alpha_2} \sin^2\beta(\alpha)\,d\alpha \quad (8)$$

The "terrain view factor" is defined as $1 - \psi_{sky}$.

$$\psi_{terrain} = 1 - \left(\frac{\alpha_2 - \alpha_1}{2\pi} + \frac{1}{2\pi} \int_{\alpha_1}^{\alpha_2} \sin^2 \beta(\alpha) \, d\alpha \right) \quad (9)$$

$\beta(\alpha)$ can be obtained from Equation 5. Its form is independent of the inclination of an edge so no additional complexity occurs with real scenarios. So what remains to be done is to accomplish the integration of $\sin^2 \beta(\alpha)$.

There is no symbolic solution for it but the calculation of the numerical solution is no problem with common math programs.

The relation between $h(\alpha)$ and $d(\alpha)$ has the most significant impact on the SVF. The point of reference that the SVF is calculated for is marked POR in Figure 7.

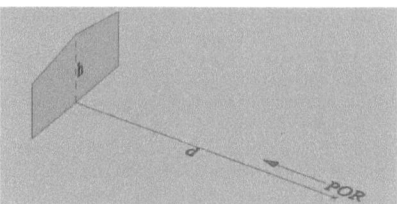

Figure 7 The h to d quotient is a critical factor for the calculation of the SVF.

The h to d quotient rises as the distance between the point of reference (POR) and the relevant building declines. Figure 8 illustrates the run of the SVF-curve for an h to d quotient starting with 0 representing a building in infinite distance up to 10 representing a building with the top edge at a vertical angle of about 85 degrees. As the h to d quotient grows also the horizontal opening angle $\alpha_2 - \alpha_1$ grows approximately at the same rate as the vertical angle.

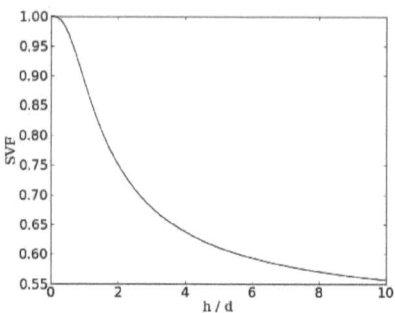

Figure 8 The SVF depends on the relation between h and d. At an h to d quotient between 3 and 5 the SVF curve starts to flatten.

From Figure 8 it can be seen that the exact calculation of the SVF is less important for h to d quotients above about 6. However for low h to d quotients also small inaccuracies can cause large errors.

EVALUATION

Whereas the performance and the accuracy of the calculation of the SVF depend on the number of slices in case of the discretization approach there is no more need for discretization in case of the functional-based approach.

Accuracy

The accuracy in case of discretization depends mainly on two parameters:

- The geometry of objects which are part of the scenario to be calculated.
- The width of the slices and the level of discretization respectively.

The calculation of the SVF of a particular slice is based on the middle value of β which is used as an average value for the height of the whole slice. So in case of straight edges in the α-β-projection there is no error at all because the middle value exactly matches the average value.

However if there is a vertex causing an offset in the projection that does not coincide with the border of a zone an error is caused. (Figure 9)

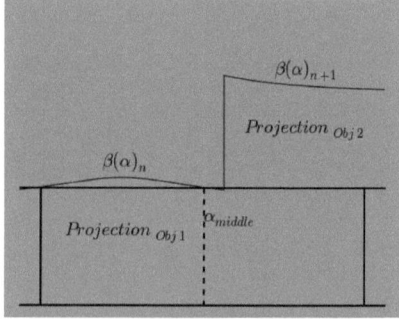

Figure 9 The middle line of a slice is close to an offset in the projection of two objects. Thus the error in the region of the slice becomes significant.

The size of the error depends on how much the average value differs from the middle value of the relevant slice. For a particular case the error was calculated for different h to d quotients.

In the following example a set of two buildings with a total opening angle of 6 degrees is investigated. Since the calculation of the SVF comprises a calculation for 360 degrees the scenario is assumed to be repeated for a complete rotation.

Table 1: Error in the calculation of the SVF for the scenario illustrated in Figure 9. There are two types of objects, one with a horizontal opening angle, $\alpha_2 - \alpha_1$, of 3.01 degrees, the other one with 2.99 degrees. The h to d quotient of the second object is kept constant at a value of 8.0. The width of the slices used to discretize the scene was chosen with 2.0 degrees.

H / D (OBJ 1)	H / D (OBJ 2)	ERROR BETWEEN FUNCTIONAL AND DISCRETIZED CALCULATION IN %
0.5	8.0	31.7
1.0	8.0	30.9
3.0	8.0	24.2
4.0	8.0	19.3
5.0	8.0	14.1
6.0	8.0	9.1
7.0	8.0	4.3
8.0	8.0	0.0

The figures in Table 1 clarify that the discretized calculation matches the exact one well for small differences between the relevant *h* to *d* quotients. Also rather large opening angles of the slices in relation to the opening angles of the buildings lead to good results if the edge of slice happens to be close to the edge of a projected building. However the size of the error remains difficult to predict.

Accordingly a high number of offsets (e.g. caused by many different buildings being projected on the hemisphere) increase the risk of slices causing an error.

To avoid for sure as unfavorable conditions as represented in the top part of Table 1 slices with an opening angle clearly below the opening angle of the relevant buildings should be chosen.

Performance

At the same time as the opening angle of the slices is decreased the number of calculation steps is increased. With the functional-based approach one calculation step is needed for every edge of the projection.

Table 2: Number of calculation steps of a complete SVF calculation (360 degrees) with the functional based approach compared with the discretization approach depending on the horizontal opening angle, $\alpha_2 - \alpha_1$, of the slices and the number of projected objects.

NUMBER OF PROJECTED OBJECTS	OPENING ANGLE OF A SLICE, $\alpha_2 - \alpha_1$	NUMBER OF CALCULATION STEPS FOR DIFFERENT CALCULATION METHODS	
		FUNCTIONAL	DISCRETIZED
15	2.0	15	180
25	2.0	25	180
25	4.0	25	90
50	2.0	50	180
100	2.0	100	180
100	4.0	100	90

Obviously the number of calculation steps for the functional based approach depends uniquely on the number of objects that are taken into regard whereas for the discretization approach it depends on the opening angle of a single slice.

For testing purposes sample scenarios were created containing a number of objects depending on a random number representing the opening angle of an object. The upper bound of the opening angle was set to 90 and the lower bound to 25 degrees resulting into 3 to 5 objects. The h / d quotients were also generated based on a random number with an upper bound according to the figures in Table 3 and a lower bound of 0. The opening angle for the slices was set to 2.0 degrees. The simulation was run 100 times with each set of parameters. Table 3 contains the average values for all 100 test runs of each test run.

Table 3: Error in the calculation of the SVF for random test scenarios. Maximum h / d values were varied between 0.5 and 15

MAXIMUM H / D	ERROR BETWEEN FUNCTIONAL AND DISCRETIZED CALCULATION IN %
0.5	1.6
1.0	2.1
3.0	2.4
4.0	2.3
5.0	2.5
6.0	2.0
8.0	2.4
10.0	2.0
15.0	1.7

CONCLUSION

A new method of calculating the SVF has been presented which is based on the calculation of the vertical angle β as function of the horizontal angle α. With the input of the geometry in a vectorial form the calculation of the function $\beta\,(\alpha)$ is possible for horizontal edges as well as for inclined ones. The geometric circumstances have been illustrated in order to set up the function that needs to be integrated to calculate the "terrain view factor" which is defined as *1-SVF*. The integral that needs to be calculated for the terrain view factor cannot be obtained in a symbolic form anymore but common math programs like Wolfram's Mathematica or MathWorks' Matlab or Python's SciPy can provide numerical solutions in real time.

Other approaches exist, like the ones presented by Gál, Rzepa, Gromek and Unger 2007 based on an urban vector database. They divide the hemisphere into slices representing equally sized parts of the hemisphere. Oke gives SVF values for commonly occurring geometric arrangements (1987), one of them a basin of constant height. The calculation based on the division of the hemisphere into slices as well as the functional based calculation use Oke's SVF value for the basin.

The calculation based on the division of the hemisphere is simple and works well for most scenarios. In some rare cases however a significant error may occur in particular if the opening angle of the slices is close to the horizontal opening angle of the surrounding objects.

With the functional based approach the exact calculation of the SVF can be guaranteed at any time.

Scalability of solutions provided for building simulation tools is substantial for the integration of a particular component within the whole architecture of a tool. With the functional based approach the computational effort is proportional to the number of objects to be taken into account. The implementation is complicated by the transformation of the Cartesian coordinate system into an α, β system. Thus straight lines are specified by a set of trigonometric functions which complicates the calculation of intersections of different objects' projections onto the hemisphere.

The computational effort for calculating the SVF with slices of the hemisphere depends on the level of discretization. A drawback of the approach is that the accuracy of the calculation grows at the same rate as the performance drops. The error depends mainly on two parameters: The dimensions of the objects to be taken into account and the level of discretization. Even though there is no fail-safe way of quantifying it, it hardly ever exceeded a remarkable degree in the random test scenarios even with an opening angle above 2 degrees.

Acknowledgements

Research funded by the Zentrum für Innovation und Technologie GmbH (ZIT) and the Federal Ministry for Transport, Innovation and Technology.

REFERENCES

Bradley A. V., Thornes J. E., Chapman L. 2001. Variation and prediction of urban canyon geometry from sky-view factor transects. Atmos. Science Letters pp 1?11 doi:10.1006/asle2001.0031.

Gál T., Rzepa M., Gromek B. and Unger J., 2007. Comparison Between Sky view factor values computed by two different methods in an urban environment. Acta Climatologica et Chorlogica, Universitatis Szegediensis, Tomus 40-41, 17-26, http://www2.sci.u-szeged.hu/eghajlattan/akta07/017-026.pdf

Johnson G. D., Watson I. D., 1984. The determination of view-factors in urban canyons. J. Clim. Appl. Meteorol. 23, 329?335.

Källblad K., 1999. A method to estimate the shading of solar radiation theory and implementation in a computer program. Proceedings of the 6th International IBPSA Conference: Building Simulation '99, International Building Performance Simulation Association, Kyoto (1999), pp. 595–601

MathWorks Matlab. http://www.mathworks.de/products/matlab/

Matuschek O., Matzarakis A., 2010. Estimation of sky view factor in complex environment as a tool for applied climatological studies. Berichte des Meteorologischen Instituts der Albert-Ludwigs-Universität Freiburg

Matzarakis A., Rutz F., Mayer H., 2007. Modelling Radiation fluxes in simple and complex environments. Application of the RayMan model. Int. J. Biometeorol. 51, 323-334

Oke, T. R., 1987. *Boundary layer climates*, Routledge, London and New York, 1987

Python Programming Language – Official Website, http://www.python.org/

Rakovec J., Zakšek K., 2012. On the proper analytical expression for the sky-view factor and the diffuse irradiation of a slope for the isotropic sky. Renewable Energy, 0960-1481 0960 1481

SciPy, package for scientific computing with Python [online] http://numpy.scipy.org/ [Accessed 4 November 2012]

Watson I. D., Johnson, G. T., 1988. Estimating person view factors from fish-eye lens photographs. Int. J. Biometeorol

Wolfram Mathematica. http://www.wolfram.com/mathematica/

I want morebooks!

Buy your books fast and straightforward online - at one of the world's fastest growing online book stores! Environmentally sound due to Print-on-Demand technologies.

Buy your books online at
www.get-morebooks.com

Kaufen Sie Ihre Bücher schnell und unkompliziert online – auf einer der am schnellsten wachsenden Buchhandelsplattformen weltweit! Dank Print-On-Demand umwelt- und ressourcenschonend produziert.

Bücher schneller online kaufen
www.morebooks.de

OmniScriptum Marketing DEU GmbH
Heinrich-Böcking-Str. 6-8
D - 66121 Saarbrücken
Telefax: +49 681 93 81 567-9

info@omniscriptum.com
www.omniscriptum.com

Printed by Books on Demand GmbH, Norderstedt / Germany